T0233262

SpringerBriefs in Public Health

Series Editor
Macey Leigh Henderson
Georgetown University Kennedy Institute of Ethics, Washington,
District of Columbia, USA

SpringerBriefs in Public Health present concise summaries of cutting-edge research and practical applications from across the entire field of public health, with contributions from medicine, bioethics, health economics, public policy, biostatistics, and sociology.

The focus of the series is to highlight current topics in public health of interest to a global audience, including health care policy; social determinants of health; health issues in developing countries; new research methods; chronic and infectious disease epidemics; and innovative health interventions.

Featuring compact volumes of 50 to 125 pages, the series covers a range of content from professional to academic. Possible volumes in the series may consist of timely reports of state-of-the art analytical techniques, reports from the field, snapshots of hot and/or emerging topics, elaborated theses, literature reviews, and in-depth case studies. Both solicited and unsolicited manuscripts are considered for publication in this series.

Briefs are published as part of Springer's eBook collection, with millions of users worldwide. In addition, Briefs are available for individual print and electronic purchase.

Briefs are characterized by fast, global electronic dissemination, standard publishing contracts, easy-to-use manuscript preparation and formatting guidelines, and expedited production schedules. We aim for publication 8-12 weeks after acceptance.

More information about this series at http://www.springer.com/series/10138

Michele Battle-Fisher

Application of Systems Thinking to Health Policy & Public Health Ethics

Public Health and Private Illness

 Springer

Michele Battle-Fisher
Wright State University
Kettering
Ohio
USA

ISSN 2192-3698   ISSN 2192-3701 (electronic)
SpringerBriefs in Public Health
ISBN 978-3-319-12202-1   ISBN 978-3-319-12203-8 (eBook)
DOI 10.1007/978-3-319-12203-8

Library of Congress Control Number: 2014954758

Springer Cham Heidelberg New York Dordrecht London

Springer is part of Springer Science+Business Media (www.springer.com)

To my Chris
To my most perfect Brandon, Michaela
and Stephen
To Mom, Dad, Stephanie and Amy
To Grammy, Paw Paw, Grandmollies,
and E-Paw

Acknowledgements

This book culminates 3 years of discovery and self-acceptance. Thank goodness, I was not alone. My life did not stop while researching and writing this book. However, the lives of those closest to me were affected. The universe brought us together. I hope that remains the case after this book. I am no longer ashamed of my abilities thanks to the persons listed below. I apologize if I missed anyone.

I would like to express my deep gratitude to my family—close, extended, and ancestral. I would like to thank the 2012 Institute on Systems Science in Health (National Institutes of Health's ISSH) and later, the Bertalanffy Center for the Study of Systems Science for validating my love of all things systems. My special thanks are extended to the faculty, staff, and students of the Wright State University Master of Public Health Program, especially Dr. James Ebert, Lori Metivier, Dr. Nikki Rogers, and Sylvia Ellison. Aaron Riley, your advocacy is worth spreading and thank you for contributing to this book.

Thanks are also due to Justin Lyon, CEO of Simudyne for his time and invaluable contribution to this book. I would also like to thank Rachel Fenercik for the sharing research that puts systems into policy practice as well as contributing with ideas and suggestions for the draft. I am particularly grateful to Dr. David Shoham of Loyola University, Chicago for his earlier comments and critique of my concentric model. Sigrid Fry-Revere, thank you for being a wonderful mentor and more.

Thanks to Mary Anne Benner of Aligif Studios for the supplying graphics of my model; to SpringerBriefs Series Book Editor, Macey Henderson, JD for the push to excellence even when it hurt; to my Springer International Public Health Book Editor Khristine Queja; to all of those who have tripped upon Orgcomplexity, please continue to do so and open it in a new tab. Finally, thanks to failure and scrapped knees. I wish we would stop meeting so often. It was not failure, but painfully necessary growth.

Ohio, USA
Michele Battle-Fisher

Contents

Part I
Systems Within Health Policy and Ethics

Chapter 1
Framing and Revisiting Ethical Policy with a Systems Perspective

We take then our point of departure from the objective fact that human acts have consequences upon others, that some of these consequences are perceived, and that their perception leads to subsequent effort to control action so as to secure some consequences and
avoid others. Following this clew, we are led to remark that the consequences are of two
kinds, those which affect the persons directly engaged in a transaction, and those which
affect others beyond those immediately concerned. In this distinction, we find the germ of
the distinction between the private and the public. (Dewey 1987)

Unfortunately, there is no universal agreement on what *is health policy*. In accordance to the World Health Organization's definition of health policy, "decisions,
plans, and actions that are undertaken to achieve specific health-care goals within a
society" constitute health policy (World Health Organization n.d.). In most general
sense, the public policy supplies a set of strategized principles that undergird laws
and statutes that are enacted by governmental bodies. There are private interests
outside the government that may be involved in policy debates and even its ratification. Some of these outside interests exist only to exert influence on all levels of
policymaking.

To the extent that legitimate political authority may allow, does public policy by
its nature hammer out social decisions for the rest of us? If a strict interpretation
of the public within the policy were taken, governance acts as the implementation
force of the policies. While governance is no less important than the act of creating
the policy measures, it is unfortunate that the appliance of governing has been rel-
egated to governmental bodies. These policies inform adherence and accommoda-
tion of social, ethical, and behavioral expectations that promote the general prosper-
ity of a society. From where do these expectations come? The policies, particularly
those involved in health welfare, cross into people's lives in the most ultimate way.
People live their well, or more often *unwell* lives, making sense of how to uphold
the strictures of the policy within the context of their private lives. For some, the
policy rests far in the background or is not known to them at all. Individuals often
reap the benefits of the policy without the active acknowledgement of the policy as
grand overseer of that social profit.

Policy governance executes prescripts at the public level, which trickle down to
a mass of private lives that are called to act upon those prescripts. While these have

© Springer International Publishing Switzerland 2015
M. Battle-Fisher, *Application of Systems Thinking to Health Policy & Public
Health Ethics,* SpringerBriefs in Public Health, DOI 10.1007/978-3-319-12203-8_1

well intentions, the activation of the policy is often uneven in this effect. Health disparities exist while well-intentioned policy was created for the very reason of eradicating them. A revised approach would require to not ignore the innate structure within the fabric of the policy. Building upon the World Health Organization's definition, health policy must better dissect its systemic elements as follows:

- Interdependent social realities
- Epidemiological goals and measures
- Policy initiatives (past, present, and future)
- Governance impacts on the healthcare system
- Health related actions of agents affected by the policy
- Ethics and morals, both individually and collectively
- Political climates

Note that this reconstruction of the health policy realizes the arc of policy as a confluence of public interests and private welfare. *Reactive policymaking* responds to an outcry that merits immediate action and governance (Torjman 2005). Policies may be made to target the short term or the long term concerns. In media, policy is often announced as an act being passed in response to a politically acute problem in hopes of reversing and/or reducing any emerging effects based on the chronicity of the health concern. The exigency of a particular situation demanding does not excise the fact that care must be taken in its development and implementation. However, network research by Crane (1991) has shown that a problem from sparse to more populated problems may more quickly spread than anticipated. Health policy has to worry about medical as well as systemic spread.

> ... if the incidence [of the problem] reaches a critical point, the process of spread [within a network] will explode. (Crane 1991)

The magnitude of the coverage of the health policy in part is defined with epidemiological evidence over time or proof of an emergent need happening now. Often in the aftermath of activating the existing policy, new developments that make evident the need for tweak or overhaul appear during the act of its use. Assessment while doing so (such as outcome evaluation) is often necessarily, a good business. System-based assessment for future policy allows for simulation of ready data to anticipate "what ifs" versus real-time trials where it is baptism by fire with no safety harness. If a policy is afforded the liberty to be worked out over time without duress, policy stakeholders may relish in the ability to break down and reassess the policies before implementing them into action. But public health always has fires to put out.

Meadows and Robinson (1985) spoke of the public's mystery of the "logic" behind how the social policy was created. What may be clear to the public is the overarching paternalistic goal of the policy. This logic of how it came in existence may not matter to the public if the policy is working. If the policy is found to be less than successful with continued ill effects on the community, the public may ask for answers from policymakers.

Systems Thinking Over Linearity

Phillip Tetlock (2005) sought to explain why political experts fail so badly at discerning complexity, let only forecasting the civic future. In order to more fully understand this phenomenon, he compared two styles of thinking. Tetlock (2005) called up Isaiah Berlin's prototypes of the fox and the hedgehog to explore patterns of political prediction in terms of discrimination of information to the calibration of that information (see Berlin 1953). The traditionalist "hedgehogs" are myopic, often resorting to the use a simple model which suits his ontology. The "foxes" are quite sly yet scattered in approaching judgment calls. Foxes rely on many mental models without discerning them any further. In the end, if you place your policy bets on the fox do so at your own risk though the odds are far more dismal with a hedgehog. Better yet, formal modeling with experience as called upon by von Bertalanffy (1962) outwits the fox and leaves the hedgehog's myopic judgments in the dust.

Tetlock is speaking primarily of getting to what is called structural knowledge (see Brehmer 1990). Changing judgments about a system with such structural knowledge can lead an organization to the operations of a model and is a requirement but it cannot discern how the system might react (Hamid 2009). How might this all affect policymaking? Where should stakeholders with the quill of governance intervene? In systems, there are leverage points. According to Meadows (1999), a leverage point is power. These leverage points "are places with a complex system where a small shift ... can produce big changes in everything" (Meadows 1999). It is great if policy can uncover a leverage to change the course as intended but it is a disaster if used selection of this leverage turns out to be incorrect (Meadows 1999).

For those who view the world as a constant interplay of its elements, this is *systems thinking*. Systems thinking is a methodological call for a critical reappraisal of policies under a new critical eye. It is an approach to understand how a whole of interrelated parts change dynamically over time. How would one know if he or she is a systems thinker? In the end, systems thinking resolves interrelationships in time and space. Booth-Sweeney and Meadows (1995) offered a laundry list of the characteristics of a systems thinker. A systems thinker would naturally:

1. See the "whole" picture
2. Change perspectives to see new leverage points to intervene
3. Look for interdependencies in elements
4. Pay attention to the long-term and not be swayed by short-term results
5. "Go wide" to see complex causes
6. Focus on structure not blame
7. Hold the tension of paradox and controversy without rushing to resolution
8. Make systems visible through maps and simulations
9. See oneself as a part of, not outside of, the system

The point is not to display a neon sign to the world that one is a "systems thinker" or exercising "systems thinking." No disclaimer must be affixed to the start of every policy document with the marque to the effect that no interdependences of the components were harmed in the development of this policy. Evidence of a systems approach would be demonstrated therein often in a way that welcomes the paradigm to the consumers of the policy in the most approachable manner. Systemic judgment that is treated as actionable recognizes interdependencies and changes in light of unforeseen developments. Policy can only look at some any systemic inputs and drivers (components) at a time. Likewise, all initial conditions are impossible to know beforehand. But a policy that is engaging systems thinking has a better fighting chance to affect and realize maturing policy changes.

Social connection is made in shared space. The goal of policy is to uncover meaning behind and within a system in which health policy exists. But is that system of any consequence to society? Systems may appear to be the "feminine" counterpart of the masculine, gold standard methods revered in health research today. But complexity is no less important and there is strength and flexibility in those loins. Ghaffar et al. (2013) wrote of "changing mindsets" and getting a place at the big kids' table of stakeholders. Systems thinking is stuck in the periphery as a mode of summarizing policy as the emergent organism that it is. Keijser et al. (2012) said that policymakers deal with "deeply uncertain problems." A policy may originate or be resuscitated from a recent catastrophe, a marked change in negative health outcomes, or even the necessity to better stay in line with the political landscape. Health policy must not underestimate the existence of "dynamically complex" situations "between systems, markets, institutions, products, regulations, actors, and policies" (see Senge 1990). Ralph Stacey (1999) said, "The less certainty and agreement had around an issue, the issue becomes more complex." The balancing act is keeping society's head above water.

What Is a System?

The definition of a system is foundational to systems thinking. Not unlike other attempts to define the physical and social worlds, there are many definitions given for "system." The father of modern General Systems, Ludwig von Bertalanffy defined a *system* as "an entity that maintains its existence through the mutual interaction of its parts to achieve" (von Bertanlaffy 1968). For starters, a system is comprised of interconnecting parts that affect the integrity of the whole when the components change over time. Systems denote interaction and flow of elements. The obvious "what ifs" are often are taken at face value, and are used to inform decision-making. The elusiveness system takes some work to discover but can be done through various systems approaches and methodologies. Until the recent advent of sophisticated computer based visualization and analysis programs (such as Vensim PLE, STELLA, iTHINK (used specifically for applications to policy), UCInet, Pajet, Linkurious, Gephi and NodeXL), social systems were described metaphorically. There were no computer based tools to mathematically formalize the theoretical hunches, or a way to reasonably concatenate and simulate this big data's complexity until most recently.

What is surprising about a social system? Ross Ashby (1956) wrote some years ago that "the harder you push, the harder the system pushes back." The hard tobacco control pushes, other drives, anticipated or blindsiding, recalibrates the tobacco system after the passage of US Public Law 111-31. Systems are built upon interaction, or as in the cases presented in this book, specifically predicated on social interaction. External environmental factors can affect how a system operates. In addition, feedbacks may also be apparent in a system. With feedback, there is a continuous flux in social influences from the external environment that require recalibration of the system. Systems thinking requires "seeing" beyond the common go-to elements of how policy has always been approached. Systems thinking can harness an understanding of social elements that often unpredictable and uncontrollable. What policymakers can control is their approach to those changing, interconnected elements.

> Systems thinking leads to another conclusion–however, waiting, shining, obvious as soon as we stop being blinded by the illusion of control. It says that there is plenty to do, of a different sort of "doing." The future can't be predicted, but it can be envisioned and brought lovingly into being. Systems can't be controlled, but they can be designed and redesigned. We can't surge forward with certainty into a world of no surprises, but we can expect surprises and learn from them and even profit from them. We can't impose our will upon a system. We can listen to what the system tells us, and discover how its properties and our values can work together to bring forth something much better than could ever be produced by our [political] will alone. (Meadows n.d.)

Accountability in evidence-based policymaking can only extend as far as the information at disposal of the policymakers. The systems used to debate and formulate health policies are based on the information actually used to make that decision. *Inputs* are components inserted into a system and affect it in some measurable way. Why is this distinction important to emphasize? The system will show characteristics that radiate from this interaction known as the property of emergence. In other words, the characteristics emerge not from components per se but from their interaction. These characteristics would not have been there without the whole result of the interacting components.

Policy "Bread Crumbs"

Policymakers have been sorely aware that the many moving parts related to a public policy were present. Harvesting and understanding those parts into a comprehensive way, emerging as a whole has been more difficult. The interplay of the components will change when people are added in the equation. The policy defines explicitly the state of the current system with intention toward change in the future. It is important to connect the concept of state of the system to policy success. It takes time to realize what is happening in a system in terms of social impact. New policies and its components are not separate from the histories left by ones before.

According to Forrester (2007), there is a life span for any component of a system. This system variable is bonded to its history ("past path") until it dies in the system.

Forrester (2007) noted that in the short term, there is less chance for error as the system. But the longer the policy is active, the game changes across the landscape. The policy is often not the last attacking a particular problem. There should ideally be a return to the policy unless outside forces torpedo it, there is a built-in shut-off mechanism, or perhaps the situation targeted by the policy no longer exists (see Bardach 1976).

> [Systems thinking and tools] give people the ability to "see" a neighbor's backyard even if that backyard is thousands of miles away. They also confer the ability to "experience" the morning after—even if the morning after is tens of thousands of years hence. (Richmond 1993)

The complexity of data constituting people, decisions, available information, and even failures is important to recognize, as historically policymaking has not embraced such complexity and systems thinking. While policy has tended to lean to look at the big picture, far less policy work has ventured into the nitty-gritty structure of these relationships across organizations and stakeholders (see Drew et al. 2011). Doug Luke (2005) advised that community-based science cannot be divorced from its social and structural context. Luke (2005) called for a *methodological consilience* litmus test. The method of explaining a social phenomenon must be more than the one that is familiar and widely accepted. In the end, do the methods we employ in health policy fully reveal and dissert the structural realities of health? (Luke 2005).

> The problems that we currently face have been stubbornly resistant to solution, particularly unilateral solution. As we are painfully discovering, there is no way to unilaterally solve the problem of carbon dioxide buildup, which is steadily and inexorably raising the temperature around the globe. The problems of crack cocaine, ozone depletion, the proliferation of nuclear armaments, world hunger, poverty and homelessness, rain forest destruction, and political self-determination also fall into the category of "resistant to unilateral solution" (Richmond 1993).

> Systems theory stands for attempting scientific interpretation and theory where previously there was none. (von Bertalanffy 1968)

The father of "systemologists," biologist Ludwig von Bertalanffy was an innovator in conceptualizing biological objects undergoing change as a system. In order to fully understand systemology of falsely disparate widgets, those widgets have social currency thus "every scientist thus living in his object." According to Bertalanffy, scientists (or policymakers) live within the system that they are observing (Pouvreau 2014). Bertalanffy and his contemporaries feared the reduction of systemology to a set of equations and maps. They feared the measurement over theorization and agency (Pouvreau 2014). Human agency must never be ignored otherwise agency will rear its defiant head like a Kraken. Researchers in social networks warn of the bias of structure over social cognates that influences the very structure measured (see Totterdell et al. 2008; Kalish and Robins 2006).

The Mental Model and Its Shortcuts

> Rigorously defining constructs, attempting to measure them, and using the most appropriate methods to estimate their magnitudes are important antidotes to casual empiricism, muddled formulations, and the erroneous conclusions we often draw from our mental models. (Sterman 2002)

> ...all decisions are made on the basis of models. Most models are in our heads. Mental models are not true and accurate images of our surroundings, but are only sets of assumptions and observations gained from experience. Mental models have great strengths, but also serious weaknesses ... computer simulation models can compensate for weaknesses in mental models. (Forrester 1994)

The book, *The Fifth Discipline*, written by Peter Senge (1990) introduced the idea of the mental model as generalization. At the core of Senge's (1990) book are the five principles of learning. These principles comprise of:

1. Personal mastery
2. Mental models
3. Building shared visions
4. Team learning
5. Systems thinking

Systems thinking is the overarching tie through these principles. Included as one of these principles, the mental model specifically is the premature and often under analyzed mental representation. Mental models are blind spots that obscure the need to dig deeper. The mental model is not the problem; failing to return to more nuanced analysis, accounting for the mental model, is. Heuristics or cognitive shortcuts that serve as social rules, which often fail and offer a false sense of security (Gigerenzer and Gaissmaier 2011). But policy must account for the past to move forward. Heuristics unwittingly act as undesirable, often detrimental noise. When told that something appears too good and simple to be true, that heuristic often is faulty logic. It is simple to take in and may in the end fool in its simplicity. Preference, according to Gigerenzer and Gaissmaier (2011), reigns supreme over measurable inferences. People do not go through the day modelling against cognitive shortcuts to decide between lattes with soy over a Frappuccino. A decision may be made quickly as work starts in 5 min, or the cappuccino machine is out of order.

When the stakes are high such is the case of the systemic effects of a health policy, formal modeling is indispensable. Vast amount of research supports the behavioral decision model (BDM) (see Hogarth 1987; Thaler 1991). In such a case of BDM, human cognitive abilities have been experimentally shown as limited. The call here is for policymakers to avail themselves the policy road maps and the social footprints left as a consequence of that governance decision. The public has short memories about the latest health scare or exploding epidemic unless a personal narrative continues to strike them. The ramifications of an unsuccessful policy linger glaringly for policymakers and this is not a time for cognitive frugality.

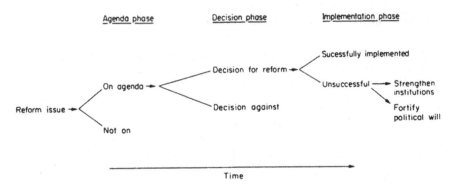

Fig. 1.1 The linear model of policy reform. (Reprinted from world development, 18 (8), Thomas, J. & Grindle, M., After the decision: implementing policy reforms in developing countries, 1163-81, 1990, with permission from Elsevier)

Jake Chapman (2004) offered the warning that reigning policy making approach is rational. Not much has changed since the Chapman's monograph was produced. But if there can ever be a way that behavior can be annulled and controlled by a policy, and then rationality would suit the purpose of policymaking. However, the policymaking approach cannot rest on reductionism. Below I introduce two opposing illustrations of policy. Figure 1.1 offers Thomas and Grindle's (1990) graphical representation of a linear approach to policy making. Next, Fig. 1.2 presents the Dynamic Adaptive Policy Pathways as one example of a complexity-based approach to policymaking (Haasnoot et al. 2013). How are the two approaches to policymaking different, or perhaps similar?

The calculus of policy making is more often than not left to experts to hammer out the dialectics. People are not classically rational; neither should our approach to policy be limited by a misplaced bias toward linearity across the board. Human rationality is gilded photo shopping. The appearance is pristine and hypnotically enticing but is improbable to achieve. To this end, Chapman (2004) offered four factors that are woefully underestimated with *rationality*. They are, according to Chapman (2004):

1. Feedbacks, which loop forward and backward in effect
2. Complexity
3. Interconnectedness (though I add that the parts are more than connected, the parts additionally interact and self-organize)
4. Globalization

One would be hard pressed to not find elements of many, if not all, of these factors at play within public policy. But Fig. 1.2 more closely fits these requirements. The very reason for policy is impact. Therefore, the wording of a policy is calculated as to hopefully reduce the risk of misinterpretation. The implementation trajectory

Fig. 1.2 The dynamic adaptive policy pathways model. The dynamic adaptive policy pathways was created to help explain the implications of the uncertainty of external factors in global sustainability policymaking. The model brings together two elements inherent to policymaking- adaptive policymaking and adaption pathways. This figure represents this systemic approach to tracking and understanding uncertainties and monitor outcomes in light of this uncertainty. (Reprinted from global environmental change, 23, Haasnoot, M., Kwakkel, J., Walker, W. & Ter Maat, J., dynamic adaptive policy pathways: a method for crafting robust decisions for a deeply uncertain world, pp 485–498, 2013, with permission from Elsevier)

taken as a result of the policy after the ink dries is not. One could say that policy has less to do directly with the outcomes based on personal actions. Policy is responsible for fostering an environment based on legally sound, agreed upon rules that serve to support decision-making. Warner (2002) pointed out that a law, or even a policy, cannot organize a public (system). The people then act within the constraints of those prescripts, for better or for worse. But if the population (public) turns out to be adversely harmed by their personal actions (private), policy is often revisited to mitigate these newly significant social developments. Was there a policy in place that should have protected to public? If so, was the scope of that policy sufficient to cover the magnitude of the event? If there was no applicable policy in place, how was the health concern handled in the absence of such guidance? If there were policies that could be employed, did they work and to what extent were they successful?

Martinez-Garcia and Hernandez-Lemus (2013) wrote of the stability of public policy in its applicability and potency in the present social environment. In other words, policy is by its nature variable in effectiveness across different segments of the public. Christopher Keane (2014), in his book *Modeling Behavior in Complex Public Health System* made the point that public health is evolutionary. Navigating health is akin to "two people playing a cooperative game" (Keane 2014). But some people break the rules in Monopoly. In that evolutionary game, personal and collective forces collide and must deal with "consumption" of limited resources (Keane 2014). In public health, there is an assumption made that there will be an "emergence of cooperation" to a common healthy end (Keane 2014). This does not always transpire.

Veterans' Health and the VA Healthcare System

When health policies are enacted, policymakers and the society at large speak of the side effects of that policy on real people. The power of harnessing support using this model is to underline the delicacy of linked relationships. Schelling (2006) warned to look away from the light of coveting constant *balance* or equilibrium. Balance is just an event, not the savior of a policy. Change is coming no matter the drive to maintain that balance. The society reacts under its mandates. Thus, it is more appropriate to view equilibrium as a goal oriented horizon point, but a result nonetheless will change again (Schelling 2006). However, unless the chain of events central to the policy disappears and becomes inconsequential, life still churns along and the residuals of policies ride on.

> Most policies fall far short of maximizing benefits. The primary reason, (Zeckhauser) would argue, is not that policymakers are ignorant or ill informed about these matters, though assuredly some of them are. Rather, policymaker's concerns, quite appropriately, go beyond choosing the policy alternative that offers the greatest level of efficiency benefits …. The political process recoils from the "let the chips fall where they may" nature of traditional efficiency maximization. (Zeckhauser 1981)

A popular vehicle for communicating a researched position to policymakers is the policy memo. Below, the following memo is meant to illustrate how systems may be used to targeting the disjointed Veterans Health Administration (VHA) system's effect on veteran health outcomes.

Case for Consideration: Veteran's Health

Subject: Understanding and Aligning Formal and Informal Structures of VA to Improve Health of the Military

The Veteran Affairs (VA) health system represents the largest integrated system in the nation. While current military are served by the Department of Defense for medical care, upon separation of duty the veteran is under the care of the VHA. The VA has is made up of 1700 facilities which are charged with the care of over 6.3 million veterans each year. A healthy military is a strong military. A healthy military helps to ensure the best quality of life of the service members, their families as well as their communities. The proud employees of the VA have answered the noble call to serve our military. As health is a partnership across agencies and patients, the VA must place a priority on real systemic successes and lapses beyond the organizational outcomes expected to succeed. The workings of VA support and hinder continuity of care. In addition, the service member and the veteran enter a web of clinics and support services with that can be dizzying to navigate and disjointed in organization. It is imperative that the Veterans Health Administration (VHA) honestly measure the true complexity of serving its patients. A lean and nimble VA system is one that does not fear the complexity before it, but rather honestly evaluates the network of layers tying medical services and patients. *Understanding and aligning the formal and informal structure of VA requires a new plan of attack, one that starts with discovering the power in looking at medical services and patients as a changing web of relationships involved in patient care.*

Honestly Assess the Complexity of the VA as a Partnership Between Patient and Services

Better health outcomes of patients require a consistency in access and quality of services. In order to understand how this can be accomplished, government by nature provides an official structure. There is official guidance and coordination. The system that you cannot see and overlook may be the more important one to understand. Why is a veteran in domiciliary care not getting connected to the hospital for her worsening PTSD (Post-traumatic Stress Disorder)? Is the counselor finding it difficult to communicate inter-agency? Is there a backlog? Did the patient just not show up and what follow up was taken and by whom to engage the veteran? Is there a gap in communication at the interagency level at the hospital? Underneath the VA official hierarchy lie a network of relationships between the patient and their services. A way of visualizing the structure of a system is to draw a network map. I have included one to bring home the importance of the informal connections between VA organizations and the patient that deviate from protocol. As demonstrated by the illustrative network below, any disruption in the organizational flow of relationships between the VA and the veteran will delay medical services. If there is no mechanism with VA to lessen the effects of systemic breakdowns such as the ones discussed, the patient and his or her care falls prey to the breakdown of his or her medical services network. These lapses may radiate from operational breaks in the formal system. But often these concerns happen with the daily grind of serving patients without an organizational chart in sight. Through policy, there must be an appropriate and appreciable response to such breakdown on both formal and informal levels. In addition, investigating informal decisions and actions on the part of VA employees gives another element of how the policy is really acted upon.

Network Figure of Possible Informal Organizational Breakdown Between the Patient and the VA

In this graph, there is a disconnection of the veteran patient (the circle) and the VA hospital (the sphere) which interferes with the medical plan required by the domiciliary. Also, there is no connection between the clinic (square) and counseling (diamond) segments in caring for the patient's PTSD although the patient (circle) is still connected to both. (Fig. 1.3)

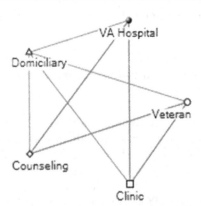

Fig. 1.3 Hypothetical VA patient care scenario. This is a network graph of possible coordinating agents of the VHA system as nodes (veteran, VA hospital, medical clinic, counseling services, domiciliary). It should be noted that a veteran's care may not require all of the agents in the graph. For the sake of illustration, the veteran in this scenario is central to the discerning breakdowns in patient care. (Network graph created by author using NodeXL- used with permission Smith et al. (2010). NodeXL: a free and open network overview, discovery and exploration add-in for Excel 2007/2010, http://nodexl.codeplex.com/ from the Social Media Research Foundation, http://www.smrfoundation.org)

Proposed Solution

Coming before the Subcommittee on Oversight and Investigations (O&I), health-care advocates expressed concerns for the "kerfuffles" linked to unnecessary and preventable deaths of military the VA pledged to protect. In that testimony, the "kerfuffle" said to be at fault was marked as disorder with the clinic system. However, dealing with something systemically is not so simple and that testimony spoke of the wide reaching effects of systemic failure. When there are lapses which increase the negative effects of these changes, business as usual will not change a system that has already left the policymakers behind. According to a US Accountability Office Report upon request of Congress in 2012, the report highlighted inconsistencies in implementation of scheduling policies, inconsistent use of an electronic scheduling linking clinics involved in a veteran's care, as well as other forms of human error (United States Accountability Office 2012). In response to this report, the VA concurred with the findings of the GAO report, setting milestones of 2013 to remedy the outstanding issues. It appears that the milestones are not being met in the best interest of serving the veterans in the VA's care. On April 9, 2014, your committee met to debate the need for a "continued assessment of delays in VA Medical Care and Preventable Veteran Deaths." Comments before the committee substantiate the need to drill down deeper and with brutal honesty the workings of the VA clinical system. Among the comments before your committee, the Assistant Inspector General in charge of oversight said:

> The lack of a common organizational Chart (formal network) for medical facilities results in confusion in assigning local responsibility for action.... Its efforts would also be aided by discussion of the best organizational structure to consistently provide quality care. The network system of organization and the accompanying motto, "all health care is local," served the VA well over the last several decades but do not standardize the organization of medical centers (Veterans Health Administration 2014).

As noted by Dr. John Daigh in the above comments before the committee, the "airline" checklist approach that "all health care is local" a part of the problem (Veterans Health Administration 2014). I argue the health care is local and whole system based. It is part of the reality that no matter the changes made to standardize the formal VA organization structure, work happens sometimes without regard of the formal structure. Reliance on technology to tie elements of the VA system, while of merit, is only a small part of the solution. The informal networks must be acknowledged as "local" in the sense that people are act as individual agents within a system. An academic study in 2007 found that the odds of mortality were higher when there was a delay in getting an appointment beyond thirty days (Prentice and Pizer 2007). This is not only an issue of backlogs which certainly must be improved greatly. Another facet of this issue is the connection between a patient and the various services required from the VHA. I propose two policy initiatives. First, stronger measures must be taken to decrease disruptions in continuity of care for VA patients. The formal organization must be reassessed from a systems perspective. This formal organization assessment must be compared to the rules and policies used by employees of the VHA. Finally and perhaps more importantly, the informal organization structures (how things are really done) must be studied accounting for the fact that systems and members of those systems attend to change over time.

It is time to see the VA patient as an integral part of the network and work to foster an environment where patients remain engaged with the appropriate services over the long term. It is acknowledged that the VHA has begun partnerships with accomplished academic institutions to pursue improving quality with the "System Redesign Framework." These partnerships at the local level develop operational decision support systems based on simulation modelling. This is evidence that VHA is dipping its toes into systems thinking. This is a welcome and much needed development that must then be taken up practically and translated into policy changes. Any policies henceforth must approach the VA "system" as a changing system using the tools and methods ties to uncovering mechanical as well as social complexity. Veterans' lives hang in the balance.

This, then, is the aim of the systems approach: looking into those organismic features of life, behavior, society; taking them seriously and not bypassing or denying them; finding conceptual tools to handle them; developing models to represent them in conceptual constructs; making these models work in the scientific ways of logical deduction, of construction of material analogues, computer simulation and so forth; and so to come to better understanding, explanation, prediction, control of what makes an organism, a psyche, or a society function. (von Bertalanffy 1967)

The take home from the memo should be that the complexity should be tackled in policy. Saying the VA mess is complex does not go far enough. The complexity is there. As will be discussed later in this book, the act of attacking one element of a system is the least productive way to deal with issues that percolate within a system. There can be a myriad of factors, both manifest and latent, that work within any system. On top of that, there are two general classes of systems:

1. Open systems
2. Closed systems

Open systems are not cut-off from the environment around them, conventionally understood as one "having import and export (of materials, energy or information)" (Checkland 1981). In other words, there are inputs that undergo change to give output. These are systems that are affected by other systems or the environment at large within recognizable overlapping contexts. The contexts could be viewed as the social equivalent of biological subsystems. On the other hand, closed systems are cut-off from the environment, cloistered away from the effects of the world.

Policy Feedbacks

There are *negativeand positive feedbacks* which are the lifeblood of a system. Martinez-Garcia and Hernandez-Lemus (2013) clarified the role of such feedbacks in terms of public policy. Accordingly, negative feedbacks "stabilize" and correct a policy, while positive feedbacks serve as a "self-amplification of trends and diffusion of new ideas…into the policy" (Martinez-Garcia and Hernandez-Lemus 2013; see Meadows 1999). The negative feedback returns policy from going away from the intended policy goal. Furthermore, Martinez-Garcia and Hernandez-Lemus (2013) defined the necessitated recalibration of policies based on social "changes" in terms of positive (reinforcing feedback) and negative (balancing feedback) loops. For instance, what positive and negative feedbacks may be hampering VHA policy success?

With systems thinking, feedback loops are just as they sound. A system maintains its overall stability by feeding depleting (as negative) or amplifying an element back into a system (as positive) whereby self-correcting and adjusting itself. According to Meadows (1999), the optimal benchmark for success that tips the policy goal and feedbacks are invisible. In fact, according to Meadows (2008), the goal is deceiving obvious. Dynamic change in a system arrives due to the interaction of positive (reinforcing) and negative (balancing) feedbacks (see Sterman 2000). As the name implied, positive reinforcing feedbacks "amplify whatever is happening in the system" as well as lead to uncontrolled growth (Sterman 2000). The negative balancing feedbacks are there to reign in the growth of the reinforcing loops (see Sterman 2000). For example, a policy is seeking to improve the stock of immunization. The policy seeks to influence rates of vaccination among preschool children within underserved populations over the next year. An inflow could be the units of the stock (net rate of the kids that got the shots) that did ultimately get fully vaccinated. The 20% increase is the goal being targeted for the stock (Meadows 1999). The goal may be research based, political motivated or pulled out of the air because that is the way it has always been done. The new policy is to accumulate more kids with shots into the policy "bathtub" and not lose many down the drain to non-immunization (see Booth-Sweeney and Sterman 2000). However, it is well accepted that there will be children that for a myriad of systemic reasons will not get fully vaccinated by school age over that time period. Those unvaccinated children deplete as an outflow the targeted policy stock of optimizing immunization. Despite the fact that healthcare systems are open to the environmental forces, Sterman (2006) advanced a portending of the error caused by the outside as well. The system by its nature is embedded in something bigger. The social environment should not be set aside as inconsequential and superfluous in reaching policy decisions. How can we possible, realistically combat policy resistance?

There is yet another implication of negative feedbacks to policy called **policy resistance**. Policy resistance are "fixes that (continue) to fail" try after try to no avail (Meadows 2008). As a part of monitoring a policy (see Fig. 1.2), reacting to instances of myopic policy resistance that counteract necessary policy changes must not fall prey to "this is the way it will go" (Sterman 2000; Haasnoot et al. 2013). There are times when the force thwarting the success of a policy is an inside job. In the situation where the negative feedbacks of a system are left unchecked, the negative feedbacks from the environment deplete the success of the policy. John Sterman (2006) in his definition of "policy resistance" addressed failures in terms of the "tendency for interventions to be defeated by the system's response to the intervention itself."

Policy Example: Australia and Antitobacco Policies

Policymakers can begin by approaching systems with the respect due to its power to uncover social mechanisms. The well regarded and "successful" Australian Plain Packaging policy illustrated internal and external systemic changes underlie even the best made and well-intended policies.

> Australia has one of the best organized, best financed, most politically savvy and well-connected anti-smoking movements in the world. They are aggressive and have been able to use the levers of power very effectively to propose and pass draconian legislation…. The implications of Australian anti-smoking activity are significant outside Australia because Australia serves as a seedbed for anti-smoking programs around the world. (Philip Morris 1992)

Australia has since implemented and had challenged progressively with preventive policies to reduce the tobacco use and mortality in that county. An ambitious harm-reducing policy, the 2012 Plain Packaging Act required that all domestic cigarettes in Australia be packed in drab packaging with prominent graphic fear appeals on the box (see Commonwealth of Australia 2012). The main effect of this policy would be to discourage the consumption of tobacco products and lessen the burden of tobacco related illness in Australia. A destabilizing effect occurred—the sale and free offers of cigarette box stickers that conceal the health warnings such as those made commercially available after the passage of Australia's Plain Packaging Act (see Commonwealth of Australia 2012; "Canberra probes free cigarette covers" 2013)? Does such plain packaging discourage smoking when the industry of cigarette skins banks its business on the fact that the cigarettes will be consumed without the gore in the customer's faces? The risk/benefit of the consumer is tied to reaction of the tobacco industry and early entry of complementary products such as the skins. Market share is based on attractiveness and utility of the good to probable consumers. The measured reaction by the tobacco company is very much short to medium term, a knee jerk reaction to strike hard and fast to end this.

Complementary products such as the skins for tobacco packs must strike quickly in order to gain a monopoly. According to Sterman (2000), the entry of the skins before other competitors may allow the utility of the good to increase as customers become more accustomed to availability of the good (either legally or on the black market) and the usefulness of the product. So how on earth can policy offset the network effects of market share? The graphics on the cigarette box serve double duty in supporting smoking cessation while in turn reinforcing the need for harm reduction through promotion messaging as "why to quit" framing. "Why to quit" framing messaging that used smokers' stories or graphic fear appeals but did not work for smokers in with no intention to quit in the next year (Davis et al. 2011). Smokers in stages of attempting to give up tobacco were found to be more swayed by the ads. While the messages studies by Davis et al. (2011) were not technically "plain packaging," the influence of personal and collective perception is highlighted as one component of the system that will only become more apparent over time. A study of tobacco policy experts conducted by Pechey et al. (2013) employed expert elicitation methods to combat uncertainty around what could happen after the Australian legislation's launch. Expert elicitation is used to help fill in the blanks when some level of quantitative accuracy is desired. According to the experts, the "plain packaging policy" should result in smoking decreasing in adults, more so in children two years are the policy was implemented. How would this be accomplished with the policy? The harm reduction warning is still there underneath a patriotic Australian flag skin purchased by the smoker.

The Australian Department of Health and Ageing later ruled that the skins were not in breach of the 2012 Plain Packaging Act as the sticker product was not sold or supplied at the time of sale of the tobacco product. Higgins et al. (2013) brought to bear the legal question on the tip of the tongue of policymakers in examining the implications of the Australian Plain Packaging Law- was there enough evidence to merit the law? A cross-sectional study conducted by Wakefield et al. (2013) shortly after the launch of the policy, plain package smokers were found to be more likely to contemplate quitting. But contemplation does not a make a policy successful or even legal? Does the lack of replicated, longitudinal evidence harm the standing of this evidence in terms of continuing to justify the policy's broad reach? (see Higgins et al. 2013). A disadvantage to the quick policy fix is that the tub may leak. Social leverages that are linked to the policy will no doubt change. There could be fewer smokers. There could be an uptick among a select cohort of smokers though this is highly unexpected. Market share of the skins could gain traction without legal restriction. Tobacco companies could justify racking up the legal fees and weaken any legal attack. The harm reduction could become the prevailing social rule over time among smokers and former smokers. But what if it does not take hold to a tipping point that the war on tobacco is being won by public health? What effect will the artifacts of e-cigarettes and skins have on realizing future public health goals? Embrace and attack wisely the unintended effects that creep up in the aftermath of a policy. Willingly accept the artifacts have taken residence, affecting the policy's effects, and act as swiftly dutifully and wisely as possible.

Individuals need, and some may qualify, deserve, medical care. There is not an infinite piggy bank to fund remediation of every issue. Often there are immediate calls for policy to act now. There are constituents with personal values and beliefs that shaped their impassioned positions that overflow the desks of legislators' in-boxes. There are special interests that can affect social mores and values. There is media spin. Policy is about striking a balance that serves the most people in the most fiscally prudent manner. Policy changes human lives. Discovering that policy sweet spot (while owning the ramifications) is the hard part. Hopefully the sweet spot will not be obscured by a patriotic skin or business as usual leaks in the VA pipes.

References

Hamid TKA (2009) Thinking in circles about obesity: applying systems thinking to weight management. Springer, New York

Ashby WR (1956) An introduction to cybernetics. Chapman and Hall Ltd, London

Bardach E (1976) Policy termination as a political process. Policy Sci 7(2):123–131

Berlin I (1953) The hedgehog and the fox. Simon & Schuster, New York

Booth Sweeney L, Meadows D (1995) The systems thinking playbook. Chelsea Green Publishing, White River Junction

Booth Sweeney L, Sterman J (2000) Bathtub Dynamics: Initial Results of a Systems Thinking Inventory. Syst Dynam Rev 16:249–294

Brehmer B (1990) Strategies in real-time, dynamic decision making. In: Hogarth R (ed) Insights in decision making: a tribute to Hillel J. Einhorn. University of Chicago Press, Chicago

Canberra probes free cigarette covers (2013) The Sydney Morning Herald. http://www.smh.com. au/national/canberra-probes-free-cigarette-covers-20130103-2c6lo.html. Accessed 17 Feb 2014

Chapman J (2004) System failure. Demos (London). http://www.demos.co.uk/files/systemfailure2.pdf. Accessed 12 Jan 2014

Checkland P (1981) Systems thinking, systems practice. Wiley, Chichester

Commonwealth of Australia. Tobacco plain packaging act. No. 148. http://www.comlaw.gov.au/ Details/C2011A00148. Accessed 17 Feb 2012

Corporate Affairs Plan: Philip Morris (Australia) Limited. (1992) University of California San Francisco Legacy Tobacco Document Library. http://legacy.library.ucsf.edu/tid/fgw48e00. Accessed 27 Feb 2014

Crane J (1991) The epidemic theory of ghettos and neighborhood effects on dropping out and teenage childbearing. Am J Sociol 96:1226–1259

Davis K, Nonnemaker J, Farrelly M, Niederdeppe J (2011) Exploring differences in smokers' perceptions of the effectiveness of cessation media messages. Tobac Contr 20:26–33

Dewey J (1987) The public and its problems. Holt and Winston Publishing, New York

Drew R, Aggleton P, Chalmers H, Wood K (2011) Using social network analysis to evaluate a complex policy network. Evaluation 17(4):383–394

Forrester J (1994) Learning through system dynamics as preparation for the 21st century. Paper presented at: systems thinking and dynamic modeling conference for K–12 education, Concord, Mass

Forrester J (2007) System dynamics- the next fifty years. Syst Dynam Rev 23(2/3):359–370

Gigerenzer G, Gaissmaier W (2011) Heuristic decision making. Annu Rev Psychol 62:451–482

Ghaffar A, Tran NT, Reddy KS, Kasonde J, Bajwa T, Ammar W, Ren M, Rottingen JA, Mills A (2013) Changing mindsets in health policy and systems research. Lancet 381:436–437

Haasnoot M, Kwakkel J, Walker W, Ter Maat J (2013) Dynamic adaptive policy pathways: a method for crafting robust decisions for a deeply uncertain world. Global Environ Change 23(2):485–498

Higgins A, Mitchell A, Munro J (2013) Australia's plain packaging of tobacco products: science and health measures in international economic law. In: Mercurio B, Ni K-J (eds) Science and technology in international economic law: balancing competing interest. Routledge, New York

Hogarth R (1987) Judgment and choice, 2nd edn. Wiley, Chichester

Kalish Y, Robins G (2006). Psychological predispositions and network structure: the relationship between individual predispositions, structural holes and network closure. Soc Netw 28:56–84

Keane C (2014) Modeling behavior in complex public health systems: simulation and games for action and evaluation. Springer Publishing, New York

Keijser B, Kwakkel J, Pruyt E (2012) How to explore and manage the future? Formal model analysis for complex issues under deep uncertainty. Proceedings of the 30th international conference of the system dynamics society. St. Gallen, Switzerland

Luke D (2005) Getting the big picture in community science: methods that capture context. Am J Community Psychol 35(3/4):185–199

Martinez-Garcia M, Hernandez-Lemus E (2013) Health systems as complex systems. Am J Oper Res 3:113–126

Meadows D (n.d.) Dancing with systems. Donella Meadows Institute Archives. http://www.donellameadows.org/archives/dancing-with-systems/. Accessed 8 June 2014

Meadows D (1999) Leverage points: places to intervene in a system. The Sustainability Institute, Hartland

Meadows D (2008) Thinking in systems: a primer. Chelsea Green Publishing Company, River Junction

Meadows D, Robinson J (1985) The electronic oracle: computer models and social decisions. Wiley, Chichester

Pechey R, Spiegelhalter D, Marteau T (2013) Impact of plain packaging of tobacco products on smoking in adults and children: an elicitation of international experts' estimates. BMC Publ Health 13:18

Pouvreau D (2014) On the history of Ludwig von Bertalanffy's "general systemology", and on its relationship to cybernetics—part II: contexts and developments of the systemological hermeneutics instigated by von Bertalanffy. Int J Gen Syst 43(2):172–245. doi:10.1080/03081079.2014.883743

Prentice J, Pizer S (2007) Delayed access to health care and mortality. Health Serv Res 42(2):644–662

Richmond B (1993) Systems thinking: critical thinking skills for the 1990s and beyond. Syst Dynam Rev 9(2):113–133

Richmond B (1997) The "thinking" in systems thinking: how can we make it easier to master. Syst Think 8(2):1–5

Schelling T (2006) Micromotives and macrobehaviors. W.W. Norton, New York

Senge P (1990) The fifth discipline. Currency Doubleday, New York

Stacey RD (1999) Strategic management and organisational dynamics: the challenge of complexity, 3rd edn. Financial Times, London

Sterman J (2000) Business dynamics- systems thinking and modeling for a complex world. McGraw Hill, Boston

Sterman J (2002) All models are wrong: reflections on becoming a systems scientist. Syst Dynam Rev 18:501–531

Sterman J (2006) Learning from evidence in a complex network. Am J Publ Health 96(3):505–514

Tetlock P (2005) Expert political judgment: how good is it? How can we know? Princeton University Press, Princeton

Thaler R (1991) Quasi-rational economics. Russell Sage Foundation, New York

Thomas J, Grindle M (1990) After the decision: implementing policy reforms in developing countries. World Dev 18(8):1163–1181

Torjman S (2005) What is policy? http://www.caledoninst.org/publications/pdf/544eng.pdf. Accessed 19 June 2014

Totterdell P, Holman D, Hukin A (2008) Social networkers: measuring and examining individual differences in propensity to connect with others. Soc Netw 30:283–296

United States General Accountability Office (2012) VA HEALTH CARE- Reliability of reported outpatient medical appointment wait times and scheduling oversight need improvement. http://www.gao.gov/assets/660/651076.pdf. Accessed 24 April 2014

Veterans Health Administration (2014) Statement of John D. Daigh, Jr, M.D., Assistant inspector general for healthcare inspections, office of the inspector general, department of veterans affairs, before committee on veterans' affairs, United States house of representatives hearing on a continued assessment of delays in VA medical care and preventable veteran death, APRIL 9, 2014. http://www.va.gov/oig/pubs/statements/VAOIG-statement-20140409-daigh.pdf. Accessed 9 April 2014

von Bertalanffy L (1962) General system theory—a critical review. Gen Syst 7:1–20

von Bertalanffy L (1967) General system theory and psychiatry—an overview. American psychiatric association. Annu Meet 176:33–46

von Bertalanffy L (1968) General system theory—foundations, development, applications. Braziller, New York

Warner M (2002) Publics and counterpublics. Publ Cult 14(1):49–90

Wakefield M, Hayes L, Durkin S, Borland R (2013) Introduction effects of the Australian plain packaging policy on adult smokers: a cross-sectional study. BMJ Open 3:e003175. doi:10.1136/bmjopen-2013-003175

World Health Organization (n.d.) Health policy. http://www.who.int/topics/health_policy/en/. Accessed 27 May 2014

Zeckhauser R (1981) Preferred policies when there is a concern for probability of adoption. J Environ Econ Manage 8(3):215–237

Chapter 2
The Public, Private, and "Stepping on Toes" in Healthcare

> If we try to eat differently from our friends it will not only be inconvenient, but we risk being regarded as cranks and hypochondriacs… It is difficult to step out of line with… peers. (Rose 1985)

At the heart of public health is the improvement of individual private lives in order to improve population-based outcomes of the most urgent need to society. Public health and private illness are political as well as biological. The dichotomy of public and private (terms I use here reservedly for the lack of better ones) is embodied. It is social. I believe that an overly stringent demarcation of the public and private is premature and fundamentally flawed. The boundaries of the public and private, I will argue, are perforated and overlap. A "public" in health is often defined as anything that is "population based." Kenneth Burke, a renowned linguist, once said there is always a "not" to a word. "Not public" is often understood as the somatic, social, and emotional experiences central to a person. These definitions of public and private are not on the surface incorrect. These definitions unfortunately do not go far and deep enough. This concept is expanded in this book as a narrative arc, deconstructing private illness embedded in publics as a dance.

> We can't control systems or figure them out. But we can dance with them! Meadows (2004)

The discussion of public versus private health is framed with a dance lesson of the "evolutionary" bachata, a partnered dance that originated in the Dominican Republic.

How to learn the basic bachata with a partner facing the other, without getting fancy (accent in fourth count)

Basically, it is "Step-together-step touch" with beats at:

Step 1—step 2—step 3 HIP (accent)

Step 5—step 6—Step 7 HIP (accent)

An earlier version of this chapter was presented at The Complex Systems Advanced Academic Workshop (CSAAW) at the Second Michigan Complexity Mini-Conference, University of Michigan- Center for the Study of Complex Systems (CSCS), on May 13, 2013.

As with any energetic bachata, two partners in close physical proximity enter an agreement to "lead and follow" in order to perform the dance. Like taking a cue from your lead partner to start movements, we also take cues from people around us. Agents in a social system do not adhere to "global lock-and-trigger" conditions in regulating their actions (Martinez-Garcia and Hernandez-Lemus 2013). To add, complex systems acknowledge the existence of many causes and effects which is radically different in approach to a "single cause causes a single effect" (Martinez-Garcia and Hernandez-Lemus 2013).

While there is some comfort in expectation of predictability of the agents' actions in physical systems, there is no such relief with social agents. People can do what they want and often do. Policies are there to guide, support, and regulate actions that are acted out in everyday life. Human agents are endowed with the ability act upon intention if the social order allows such freedom. In order to have intention, there must be cognition. For the sake of this book, a social agent may be thought of as:

> A natural or artificial entity with sufficient behavioral plasticity to persist in its medium by responding to recurrent perturbation within that medium so as to maintain its organization. (Goldspink 2000)

Partner dancing requires "real-time coordination between a human leader and follower," and resembles other decentralized systems with "supervisory control and coordination of agent teams" (Gentry and Feron 2004). Each partner is an agent who must take physical and verbal cues to move together. More often than not, there are glitches mingled among the passion to work together. The yearly homecoming of mandatory influenza vaccinations is a good example of the difficulties in reconciling and satisfying both private and public interests. One person gets the vaccination, perhaps under his own volition or social pressure (e.g., my boss tells me to comply for the safety of patients). One person may be left out of the loop in learning of the directive to get that mandatory flu shot (organizational lapse). Another may decide to forego the advice to immunize and take the chance of illness and being singled out as a result. In the end, the dancers in this public health situation become a small public where each must abide by the rules of gestures of bachata (getting the shot). Bachata is different from the waltz. There are rules in place to guide the anticipated actions of abiding by the requirements to obtain the flu vaccination. When that gesture is something like getting that flu shot before reporting to work, some may decide to become wallflowers and ignore the requirement altogether.

In essence, there is a give-and-take among all dancers (social agents) toward a collective end result. This is not unlike the give-and-take of navigating health as a person that is a part of a public. How could this be? While living as a social creature may lack the lyrical musicality of a bachata song, effective public health action requires that the private sphere be under the watch of the public for the sake of the society as a whole. However unlike the Bachata dance and song that has a definite outro, the political intensity and prescriptive requirements surrounding a policy typically hit a loop section. The song never ends.

One partner cannot do waltz steps while the other performs Dominican bachata. In other words, an entity such as a government or even a bachata partner should not demand to have a say in his status. But this would not make sense for dancing bachata, which is by design a partner dance. Likewise, when we evoke the idea of public health, the idea of intervention into the *private sphere* is central to improving the health of the larger public. Some of our private has to give for the sake of the public.

People tend to gravitate homophilously both politically and culturally (see Aristotle 1991; see Kilduff and Tsai 2003). Michael Warner (2002) wrote a piece in *Public Culture* which later became a book of the same name, *Publics and Counterpublics*. The public is "everyone within a field of study" (Warner 2002). The public is the study population. Warner (2002) presented the argument that being a member of a public requires that each citizen participate at some level. Not being fully activated in deliberation may not be feasible. Some may not be activated to participate at all. Sometimes a citizen does not act as the hand is forced. Ethical issues are off of the radar and lack saliency, until the acute issue strikes them in the face or knocks them off of their feet (Nordgren and Morris McDonnell 2011).

Gigerenzer (2010) said that social pressure to comply with the peer network is based on intuition. The legal scholar Cass Sunstein (2005) asserted that moral heuristics are also susceptible to error. As morals and ethics are inherent to policy-making, the ill effects of such heuristics can bleed into ill-advised policy decisions (Sunstein 2005). A heuristic by nature is an artifact based on socially accepted fact (Sunstein 2005). A heuristic is not unlike pulling off that decisional Band-Aid as quickly and painlessly as possible so you can hop back on your bike. Likewise, making strategic and ethical choices does not cleanly translate to an assurance of socially or medically appropriate actions by the public under the stricture of the policy (see Adler 2005). Choice is a test of feedback based in part on ethical fallibility in following socially agreed upon ethical rules.

The Will of the Public

Aristotle (1991) tendered a guide on how to live with other flawed people in a reflective, public space. In his work, *On Rhetoric*, Aristotle (1991) said what is unavoidable is to be a part of a larger public which is composed of people with moral blemishes. In Book 1, Aristotle gives character a new name, *ethos*. This *ethos* is born of a sense of justice. But Aristotle made it clear that *ethos* cannot be mandated by law. Hannah Arendt (1958), in *The Human Condition*, said that people lapse into a social world. Because humans are so intertwined, the public realm is infinite so should be our concern for others. Our identity becomes one of the collective (Arendt 1958).

Van Kleef et al. (2010) were interested in this very question in mortals by saying that emotions should be treated as bits of information that must be understood. In their empirical findings, during a situation that requires competition, people draw off of trying to analytical weigh the emotions of the opponent (Van Kleef et al. 2010). In those situations where we are "in this together," emotions rules by getting into someone's head (Van Kleef et al. 2010). For health policy, there may be longer peace between the social and related ethical upheavals but policy revisits at some future point to restore order (see Martinez-Garcia and Hernandez-Lemus 2013). The very basis of public health rests on overlapping influences and relationships of established norms, political whims, ethical standards, and social connections (Gregson et al. 2001; see Griswold et al. 2013). As ethics and actions are both moving targets, so must be our understanding of the nuts and bolts of the system.

Sobering realization of progress… (DuBois 1965)

Over himself, over his own body and mind, the individual is sovereign. (Mill 1999)

The will of society, with our funny little orb of health hovering and bobbing within it, aligns religiously with the will of the majority. Mill (1999) wrote *On Liberty* in 1895 as a treatise on the overlay of personal responsibility and an authoritative state. He advocated a balance of Pareto efficiency as good for the majority, which is fundamental to public health) and personal autonomy (with the translation you cannot make me do it for my own good). For the sake of this argument, Mill also defends the existence of protection against prevailing feelings and opinions (Mill 1999). People, according to Mill (1999), base decisions on "personal preference." If health invokes the power to infringe on the personal liberty of an individual, Mill (1999) said that the public only have that right when the burden of proof suggests that "preventing harm to others" is the only justification. Mill also defends the existence of protection against prevailing feelings and opinions (Mill 1999). The explanations that we view as valid or potentially valid are at the mercy of the "denial of usefulness" (Mill 1999).

The need for an intellectual irritant is often necessary to spur innovative discussion. Mill (1999) wrote that "men are not more zealous for truth than they often are for error." Human nature requires adjustments to our reasoning out of realized and accepted necessity. Individuals do not expend unnecessary energy when they are comfortable with the status quo. Mill (1999) explains our propensity toward remedy as being tied to two factors: the direction of the sentiment (as in complexity versus convention) and the degree of interest in that sentiment. Otherwise, we are indifferent or opposed to seeking out alternatives.

Personal Liberty and Social Utility

What is made quite evident is the warring of the ideals of maintaining personal liberty with the maximization of social utility. What is the acceptable tipping point before we topple too far to the side of autonomy while undermining social welfare? Dworkin (1972) in response to Mill wrote that paternalism "will always involve limitations on the liberty of some individuals in their own interest but it may also extend to interferences with the liberty of parties whose interests are not in question." Mill sets an unreasonably high threshold for achieving paternalism. This certainly is not unusual with such sweeping pronouncements.

In light of the impossible achievability of Mill's requirements, Dworkin (1972) said that what is at work in reality is impure paternalism. As impure paternalism is followed, a disenfranchised class' needs are met by way of subjugating the requirements of an unaffected class (Dworkin 1972). Someone has to give up liberty so others can gain. But who wants to be the "perceived" loser? If the argument remains as a question of personal liberty, yes, someone will shoot craps. It is getting over the immediacy of the personal loss of liberty that sting.

Public health as a discipline, science, or even branch of medicine cannot guarantee that for each loss of liberty there will be a recognizable, visceral gain in health status. People live by the concreteness of their own experiences. The concreteness of a relative risk measures disease not liberty. Unfortunately, this is often to the disadvantage of public health to get buy-in. But it is perhaps the impurity of human experience that makes the work of public health the most noble of all. It is just so darn hard to lose while gaining.

References

Adler M (2005) Cognitivism, controversy and moral heuristics (open peer commentary to Sunstein). Behav Brain Sci 28:524–542

Arendt H. (1958) The human condition. The University of Chicago Press, Chicago

Aristotle (1991). On rhetoric. A theory of civic discourse. Translated by Kennedy G. Oxford University Press, New York

DuBois, WEB (1965) The souls of black folk. In: Franklin JH (ed) Three negro classics. Avon, New York

Dworkin G (1972) Paternalism. Monist 56(1):64–84

Gentry S, Feron E (2004). Musicality experiments in lead and follow dance. In: Systems, man and cybernetics, 2004 IEEE international conference 1:984–988

Gigerenzer G (2010) Moral satisficing: rethinking moral behavior as bounded rationality. Top Cogn Sci 2:528–54

Goldspink C (2000) Modelling social systems as complex: towards a social simulation meta-model. J Artif Soc Simul 3(2). http://jasss.soc.surrey.ac.uk/3/2/1.html. Accessed 1 Feb 2014

Gregson J, Foerster S, Orr R, Jones L, Benedict J, Clark B et al (2001) System, environmental, and policy changes: using the social-ecological model as a framework for evaluating nutrition education and social marketing programs with low-income audiences. J Nutr Educ 33(Suppl 1):S4–15. doi:10.1016/S1499-4046(06)60065-1

Griswold KS, Lesko SE, Westfall JM, Folsom Group (2013) Communities of solution: partnerships for population health. J Am Board Fam Med 26(3):232–238. doi:10.3122/jabfm.2013.03.130

Kilduff M, Tsai W (2003) Social networks and organizations. Sage, Los Angeles

Martinez-Garcia M, Hernandez-Lemus E (2013) Health systems as complex systems. Am J Oper Res 3:113–26

Meadows D (2004) Dancing with systems. Donella Meadows Institute Archives. http://www.donellameadows.org/archives/dancing-with-systems/. Accessed on 8 June 2014

Mill JS (1999) On liberty. Longman, Roberts and Green, London

Nordgren L, Morris McDonnell M-H (2011) The scope-severity paradox: why doing more harm is judged to be less harmful. Soc Psychol Personal Sci 2(1):97–102

Rose G (1985) Sick individuals and sick populations. Int J Epidemiol 14:32–38

Sunstein C (2005) Moral heuristics. Behav Brain Sci 28:531–573

Van Kleef G, De Dreu D, Manstead A (2010). An interpersonal approach to emotion in social decision making: the emotions as social information. Adv Exp Soc Psychol 42:45–96

Warner M (2002) Publics and counterpublics. Public Cult 14(1):49–90

Chapter 3
The Menagerie of Social Agents: People and Their Connections

There was a critically acclaimed Broadway memory play written by Tennessee Williams called *The Glass Menagerie*. The mother named Amanda, an aging, fading debutante, who longs for her lost youth, was deserted by her husband long ago and is fixated on getting her daughter Laura married. Laura Wingfield suffers from physical infirmity as well as self-imposed social seclusion. A glass menagerie is a small collection of glass figures that is zealously worshipped by Laura. The glass menagerie signified the emotional and social fragility of the main characters. The glass figures could never substitute for human interaction. The brother and narrator, Tom, is the family broker to the outside world and commits to help his mom find a suitor for his painfully shy sister. Artificial worlds are created by Laura and her family. The Wingfields create an impenetrable group based upon despondency and domiciliary seclusion. Such a distinct delineation has at its heart insiders and outsiders (Borgatti and Halgin 2011). I would argue that the family based on strong kinship would be more than a group. They form a network. Not unlike the stale air circulated in a land of bland cubicles, no new air of social reality enters the home of the Wingfields. Any tumult in their secluded world collapses it.

The brief introduction of the unavailable suitor, Jim, is the last straw for the health of the Wingfield network. Jim, who could have served as a bridge across the structural hole between the Wingfields and the rest of the world, would leave and never return. Unbeknownst to Tom, Jim is engaged. There is a fight and then a fracturing of the Wingfield clan.

> Oh, Laura, Laura, I tried to leave you behind me, but I am more faithful than I intended to be. (Tom Wingfield, scene VII)

Tom would eventually renege on this assurance from scene VII. Tom leaves Laura…and Amanda. Then there are two, Amanda and Laura, left in that dank rundown apartment with a broken glass unicorn. Each member of the family, as members of this cloistered network, brings his or her own ethics, actions, and perceptions that when taken together adversely affect the family dynamic. Closed systems in that sense are stifling. Looking at social engagement with the strictness of network formation alone is also stifling as well.

© Springer International Publishing Switzerland 2015
M. Battle-Fisher, *Application of Systems Thinking to Health Policy & Public Health Ethics*, SpringerBriefs in Public Health, DOI 10.1007/978-3-319-12203-8_3

Having a dynamic view of social relationships requires opening up to the fact that "each network will have its own structure and its own implications for the nodes (people) involved" (Borgatti and Halgin, 2011). The translation is that people and their resources flow (evolve) over time. Kilduff and Tsai (2003) noted that social networks help to peel away at the idea of embeddedness, specifying how our social interactions (transactions) overlap to the network structures. If one is to ascribe to a systems way of approaching societal change, the structure of a system is all about the changes that happen over time. While policymakers may attempt to attack singular elements of a system (which is the least likely attack to succeed with a system), it becomes apparent over retrospection that "the system's responses to outside forces is characteristic of itself" and must be engaged when revisiting policy (Meadows 2008). For policymakers, the scuffle becomes one based on leveraging expertise and political expediency to make marks on policy. It may take more than helping Amanda Wingfield to change the state of affairs.

Social Network Theory and Analysis

Social network analysis uncovers the influence of linked relationship of members of a connected network (Hanneman and Riddle 2005). Social networks celebrate the structure and agency both of which may not be obvious until the network is approached aerially (Kilduff and Tsai 2003). Networks are mathematical models of congregation of people, a heap of words, or solitary-but-social lemurs balling up in a crook of a tree. As well said by Burt (2004), networks supply context, with the clarification that only the members of the network can act upon that system of social connection until they are let in (Burt 2004). Paranyuskin (2010) states what a *social network* is brilliantly, as a shared "occasion of experience." To conjoin Burt and Parayshkin, policy accounts for serving collective health by joining together with often divergent partners as the means to a grand, bold end of improved health outcomes.

> We are concerned with social networks and what passes through these networks—friend-
> ships, love, money, ideas and even disease. (Kadushin 2012).

According to Kadushin (2012), a social network is a "set of relations between objects which could be people, organizations, nations...." Below, a list of basic network terms is provided. It should be noted that social network analysis can use a number of terms that describe the same phenomenon. For simplicity, the provided terms indicate the terms that will be used in this text. For a more exhaustive explanation of social network terms and analysis, Hanneman and Riddle (2005) is suggested for individuals that are new to the discipline. Hawe et al. (2004) provide a glossary of social network terms presented in terms of applications of social network data and analysis of concern specifically to health research.

Basic network terms (Wasserman and Faust 1994; Kadushin 2012; Hanneman and Riddle 2005)

1. Social network—"set of relationships" that contains a set of objects (nodes); mathematical models of social connection
2. Ego—The main subject in the network that is of interest if desiring a closer look at a network of that particular subject
3. Nodes (or also called Actors)—a set of objects; the individuals connected and illustrated in the graph (including the ego)
4. Alter—person tied to main subject (ego) in an egocentric network
5. Ties—links between nodes which demonstrate the strength and/or directionality of perceived relationship(s) in the network
6. Directed network—relationship between two nodes with indicating flow of information or sharing
7. Simple network—relationships visualized only by presence of a connection (no directionality or flow indicated)
8. Symmetric network—relationships of nodes that are all connected to the other nodes; everyone gives and receives from everyone else through ties
9. Asymmetric network—relationship of nodes that is not mutual across all nodes; some nodes are no connected to others
10. Propinquity—"being in the same place at the same time"
11. Egocentric network: small personal universes connected by one main subject (ego)
12. Complete network: big picture structure with no central figure (ego)

Social network analysis accounts for the interdependence of the elements (nodes) of the network. How can individuals be equated as independent if what one is trying to understand are their mathematical and social relationships? The key to social network analysis is defining which nodes to include by specifying a boundary that includes only the nodes in the analysis. There is no natural *boundary* to a network; the boundary must be rationally defined. Agreeing on a boundary specification is central to the success of application of social networks. A boundary is set in order to frame or conceptualize what people are relevant to the network. In the end, this chapter is most interested in how people are tied socially in which characteristics of prestige and power determined by the structural analysis of the network. According to Trotter (1999), the existence of a boundary is defined by the rules of exit and entry. However, systems theory calls for more intricate phenomenological examinations of "when and where (one) enters" and exits relationships (Cooper 1988). Laumann et al. (1983) offered four domains of boundary specification for social networks specifically:

1. Positional—based on membership such as the staff of RAND, employees of a certain governmental agency, all students in a particular psychology 100 class at a selected Midwestern college, or all patients of a selected medical practice
2. Event based—same event at the same time, such as all attendees at the Super Bowl or people waiting for a NYC subway at 5 am on a Monday
3. Relational—based on social relationship such as kinship ties in a Somali clan in Minneapolis or a social support network of a chronically ill patient
4. Reputational—person of influence names the roster, such The Chair on the Ways and Means Committee names knowledge experts in US budgetary issues

Egocentric is great in terms of drilling down to the structural narrative of one person and those connections. With the complete network, the intimacy is lost by broadening the conceptual lens by analyzing the network as a whole. Newman (2003) notes the futility of graphically representing overly large populations because the graph will never sufficiently represent every node that could be in it. This is often a rare opportunity, for this works not unlike a census.

Strogatz (2001) acknowledges that the structure and function of a system cannot be divorced. According to Newman (2003), there are three central tenets to approaching social networks:

1. Uncover the measurement and nature of networks statistically
2. Create models to understand these inherent relationships
3. "Prediction" of an outcome variable, based on the discrete mathematical rules of social network analysis (Newman 2003)

The key assumptions of social network theory are:

1. The aggregate influence of the group is more important.
2. The analysis is at the level of the network.
3. Patterns of relationship exist across the whole network.
4. Social networks are linked relationships (Kadushin 2012).

Borgatti and Halgin (2011) made a necessary peculiarity of network theory and/or the "antecedents of network properties" (see Borgatti and Halgin 2011). This transformation of composition can be viewed as a reorganization or emergence of a complex network, in the sense that this form of complexity is inherently on a dynamic continuum with continuous reorganization of the nodes (Halley and Winkler 2008). Networks have much more than a common interest in keeping nodes (people) together. Before employing social networks to an issue, whether the systems thinking paradigm, more specifically, social network applies must be questioned. Based on Borgatti et al.'s (2009) four *categories of network arguments* (2009) if any of these queries apply in the development of a policy, the social network perspective should be investigated further (Table 3.1).

For guidance, Borgatti et al. (2009) offered four categories for network arguments:

1. Transmission
2. Adaptation
3. Binding
4. Exclusion

Being connected takes work but working to stay connected may be harder when the person is on the social fringes. Social capital, or philo, as defined by Krackhardt (1992) requires three attributes: time, affection, and interaction. The strength of a tie is measured by a "combination of the amount of time, the emotional intensity, the intimacy, and the reciprocal services that characterize the tie" (Granovetter 1973). With the "navigability of strong ties" proposed by White and Houseman (2002), the nature of strengthened engagement fortifies within a network trust among its

Table 3.1 Application of the categories of network arguments to health policy development

Category of network argument	Policy questions to pose
Transmission	Do social elements targeted by the policy involve flow of information among people?
	Are there certain people in a network with the advantage of accessing social resources? Are some people disadvantaged in accessing social resources?
	Must the policy account for social structural changes that create different circumstances?
Adaptation	Is the basis of the health concern the issue of people making the same choices due to similar network positions?
	Is a person's position in a system possibly tied in part to similar life chances and circumstances?
Binding	Do people act as one, sharing actions and outcomes?
	Is the influence to act collectively possibly linked to structural issues?
Exclusion	Does the policy seek to influence overcoming breakdowns and exclusionary situations that hamper a person's access to social relations and health based resources?
	Do social structures change under certain social and/or poilitical circumstances? (category of transmission)
	Is the policy involved in the altering of choices when the actors are exposed to similar social constraints?

members (White and Houseman 2002). Carpenter et al. (2003) found in a simulation model of health political networks that increased burden of involvement in the affairs of the networks increased reliance on strong ties. In addition, strong ties must be cultivated to maintain such networks, with five times the maintenance effort over weak ties (Carpenter et al. 2003). Borgatti and Halgin (2011) present the distinction of a state tie from event tie. State ties have "open-ended persistence" and get at the idea of social connection. On the other hand, event ties can be bean counted. Tom and Jim, the paramour thus christened by Tom, had a flow relationship in which Jim had as the role-based state-tie relationships of coworkers and minor friends. Borgatti and Halgin (2011) define this flow as in terms of exchange. How in the world do the fictitious Wingfields offer lessons in systemic engagement? The ties in a network are rather enduring. According to Borgatti and Halgin (2011), such a persistent tie is known as a state-tie. As with the nature of the ties that will be described in this book, a state-tie "flows" in a manner that can following some rules of engagement. According to Borgatti and Halgin (2011), the three tie characteristics are:

1. Tie strength,
2. Intensity, and
3. Time duration.

People who do better are somehow better connected and leverage those connections as optimally as possible. Holding a better position in the structure of these exchanges can be an asset in its own right. This asset is social capital (Burt 2005).

Risk and Network Analysis: The Case of HIV + Risk and Homelessness

An unfortunate consequence of homelessness is a greater burden of diseases hampered by the lack of stable housing. The HIV prevalence rate among the homeless is at an alarming rate in comparison to the general population, with infection rates from five to ten times higher in comparison to the stably housed population (Milloy et al. 2012). Being homeless in urban centers has statistically been found to be a predictor of HIV prevalence (Dennings and DiNenno 2010). Positive HIV status may push a person into the societal fringes, resulting in the loss of a stable housing environment and its possibility of social support. Conversely, after a seronegative individual becomes homeless, there are concerns of exposure to risky behaviors such as needle sharing, victimization, drug abuse, and seropositive sex partners, which are all risk factors for HIV infection (Kidder et al. 2007). The policy target toward reducing the odds of contracting HIV has been securing housing and rightly so. Though the staggering numbers would support the need to take a fresh approach to interventions with homeless populations, very little research has been undertaken using social networks and system science and its usefulness to HIV policy.

Network analysis conducted by Green et al. (2013) framed the nature of social support among homeless men more than a resource that a person is given, but also as connected to social connection over time. Among a select population of homeless men on Skid Row, this study found that chronic, long-term adult homelessness was tied to friending riskier connections and also experienced increased probability of fragmented, unstable networks (Green et al. 2013). Homeless teens have received recent attention including the risk of adverse health outcomes connected to homelessness. The pressing policy issue of HIV risk-taking behavior among homeless teens to take home this point of social bond (de)construction while homeless. Within social networks, social network composition among teens may vary from that of adults. While Green et al.'s (2013) study found that fragmentation of networks was common for the chronic adult homelessness, the social support networks tend to gravitate towards the need to clique for teens. Lack of consistent housing has been statistically proven to pose grave threats to a homeless teen's health status. The immediacy of basic needs often calls upon the solicitation of resources from the homeless teen's social network, whether safety net-based or borne out of unfortunate circumstance.

Homogeneity in a social network is a measure of the skew of nodes toward a singular and identified social relationship (Kadushin 2012). Older adults tend to report homogeneous networks overwhelming composed of family. As a result, adult children bear the grunt of the burden in their parent's care (Fingerman and Birdett 2003; Umberson et al. 2010). Younger adults were more apt to name fewer relatives as close or may name familial ties as problematic (Fingerman and Birdett 2003). Children rely on parents with a marked shift in adolescence to seek support and refuge outside of the home with friends (Umberson et al 2010). Teens who leave stable housing for the streets formulate a new environment of state-ties out of necessity and survival (Mizuno et al. 2003). For homeless teens, they could build *survival networks*. In such a network, the refuted model of the risky person is not resurrected. The once predominate *risky person model* placed culpability for risky behaviors on character traits and failings of the homeless individuals that were brought with them into the homeless context (Aidala et al. 2005). Risk is instead found to be tied to the context of risk based on homelessness of which the teens navigate together (Aidala et al 2005). Network analysis can examine more widely the macroeffects of housing on a complete network, one that gives an overview of the whole structure rather than concentrating on the structural experience of one node in the network. This actuality of a survival network is based on the existence of a risky shared space of homelessness egocentrically.

Survival networks most likely would not have occurred with their current content or in its present configurations without the factor of homelessness that brought the nodes together. Friends made on the street may not have shared homerooms or civic activities before with the newly homeless teen. Lack of secure housing diminishes social capital often taken for granted by those who are stably housed. The lack of a roof over the teen's head does not open a sea of opportunity but rather lowers a glass box over the survival networks that become further removed from society. Sometimes those in health policy may wish to merely bean count the uniplex and multiplex roles in question (Barnes 1972). But those designations may not fully serve the dynamic requirements of developing health policy. An understanding of the underlying structural relationships and the social environments in which people are embedded should be a part of the policymaking process. For others, the roles become muddled with roles that support health and others that result from non-medical reasons. Barnes (1972) defined multiplexity in terms of a "multiplexity of interests" that bond the two nodes together. The tie that is formed in such instances, called a multiplex tie, tends to be very strong and enduring. According to Kilduff and Tsai (2003), this is easier said than done for a person who serves many pivotal roles to say goodbye to those relational responsibilities. Notions of roles and allegiance to the patient versus own taking care of one's own life breeds complexity at every turn.

The context of the social network matters. The focus theory approach, as presented by Feld (1981), frames social context in terms of "a group's activities are organized by a particular focus...and two individuals that share the same focus are most likely to share joint activities." To bring social context into focus for homelessness, the context becomes a pairing of identity (as in homophily) and social engagement based on that shared focus. Sharing a recognized characteristic denotes homophily (Kadushin 2012). But the overwhelming evidence of the gravity of homelessness on health outcomes would call for a broadening to the connection of elements of bond formation. When the teens come together within a network, an acknowledgement of association is presumed. Homophily does not scratch the surface on issues of social context. Homophily is not concerned with the "opportunity to meet" or ties emerging from shared context (Feld 1981).

The risk, as supported by recent network analysis of homeless teens' HIV risk then is based on position within a network and not solely risk based on personal failings (Rice et al 2012). A place where a person lands in a network may not be same as the role that person pays in a network. In addition to highlighting the implications of position in a network, the teen also must comply and conform. The Solomon Asch conformity experiments from the 1950s investigated the propensity "minority of one [to conform] versus [the influence of] a unanimous majority" (Asch 1956). There has been general agreement since Asch (1956) that a single individual is apt to conform even when the decision runs contrary to established fact. The will of the majority (public) may possibly lead to the denial of private judgment held to the contrary (Asch 1956). This denial for the sake of fitting into the survival network is conformity. Information is only so if it is novel to the network (Rapoport 1953). Rapoport (1953) found in experimental studies that an inner circle of "knowers" may slow innovation. That inner circle may be stalwart in their beliefs (risky behavior). So in the end, there is a naturalization path for outsiders to come in the core. In terms of randomness of distribution, new teens from the periphery and old, stalwart core teens intermingle. But once the outsiders are there, they may take on the difficult task of being an innovator or disappearing into the scenery and conform. The quandary here is that the policy wants to work against the propensity to conform to norms putting the homeless teens at risk for HIV if transitioning into the core.

Taking into account the unique *risky context* of homelessness, *survival networks* can be viewed in the following ways:

1. *Networks* with a socially restricting glass box that restricts the possibilities of fully realized social integration
2. *Propinquity* that may affect the state-tie composition of nodes that are accessible and acted upon by a particular teen (whether under own agency or under pressure to conform)
3. *Bond forming* that may intensify due interdependence linked to the social context (such as homelessness)
4. *Bond forming* based on common foci that may entrench relationships and increased risk of social conformity

In social networks, there may be a significant difference in the core or the periphery partitions of a social network. According to Borgatti and Everett (1999), there are three intuitive notions of the core to the periphery dichotomy:

1. A group that is built of nodes that are incident to different number of ties (highly integrated or largely distant and isolated)
2. Two groupings of nodes that form one larger group, one as core and one as periphery, where the "the character of ties within the periphery as well as within the core [are specified]"
3. "Cloud of points in Euclidean space" where the big ball of spaghetti in the center of the network are the core

Nodes in the periphery by definition are sparsely interconnected among themselves while the core forms a clique (Borgatti and Everett 1999; Persitz 2010). The core is right at the "center to the action" (Borgatti and Everett 1999).

As a result of Rice et al.'s (2012) study of HIV risk behavior, homeless adolescents were located within the core, were more likely to be female and were more likely to have been homeless for at least 2 years. The longer the teen, particularly for the young woman, is outside of the family unit, the more teens form strong, compact ties with a new definition of family. Surprisingly, being on the periphery of the tight core of homeless teens may be protective against HIV risk taking. Teens on the periphery also reported being homeless for shorter periods and are more likely to have maintained prosocial connections to home (Rice et al 2012). The solitary existence of the ill-defined netherworld between the former "home" and the tight clique of homeless teens may become unbearable for the peripheral teen to maintain. The homeless teen who is outside of the support of other teens and cannot go to a stable housing environment may in the end succumb to the human need for safety and social connection on the streets. A focus is a "social, psychological, legal, or physical entity around which joint activities are organized" (Feld 1981). According to Feld (1981), a focus serves as a social glue, keeping nodes engaged with each other as long as commonalities are fostered. Once the identity of being homeless becomes the relationship-based focus, the shared identity as well as engaging in purposeful interaction can lead to more time spent together (Feld 1981).

Highly connected, dense cores in networks are conducive due to their central location, galvanizing and passing information within that group due to the short path to get from one node to another (Borgatti 2005; Persitz 2010). However, a dense group may be more difficult to infiltrate, diminishing informational flow and hampering positive bond formations. Preliminary evidence has shown that it is not enough to be connected to a person that engages in risky behaviors. Being a member of the core group where social learning may be more likely increased the risk of the hazard of homeless teens engaging in risky HIV behaviors (Rice et al 2012). This result is in stark contrast to previous research that has found that deep social integration is often positive among teens (see Allen et al. 2005; Valente et al. 2003).

When evaluating health economics, Shiell et al. (2008) noted along with uncovering change in the overall social systems, the movement of the elements of the systems must be understood as well. They are least invested so how invested would that expert be in using its scare resource of political clout to bring harmony to the support? According to "structural holes," the node is brokering a relationship with someone that person does not already have a relationship, across a social gap or hole. Burt (2004) trumpeted the idea of "vision advantage," where the two disconnected networks that are spanned across a hole can help to bring in new ideas to the other. Long et al. (2013) contend that such vision advantage can lead to overdepletion of resources and a resulting loss of acuity that was desired in the first place.

However, node and tie failures are not ideal. If a node is removed from the network, it is called a vertex cut. In an analysis of peer-reviewed systems literature conducted by Long et al. (2013), this brokerage as a bridge between groups that were not previously connected is central to understanding whether "good ideas" as resources are being diffused. According to Long et al. (2013), the take-home messages are worth a long gander to frame approaching that next framing of homeless policy. Long et al.'s (2013) findings were:

1. The more comprising the network, the more efficiently information gets around among its membership.
2. Boundaries spanning a "hole" between the core and the periphery by and bringing in new people that has not been a part of the in-crowd (core) may not be the most reliable method of getting good ideas and positive support for positive sexual choices.
3. Teens, as brokers with their increased yoke of responsibility may just burn out and leave, creating a cut vertex.

Granovetter (1973) defines a bridge as a person in a position of linking people who would not have otherwise been connected. Granovetter (1973) found that, in general, it was the existence of weak ties in networks that had the more positive influence on behavior change as there was no competition for scarce resources. It is not assumed that every tie in a network would be designated as weak. Strong ties link friends who make more of an investment into maintaining the relationship which may also result in a more lasting relationship (see Krackhardt 1992).

If the public policy being developed pertains directly to HIV prevention, perhaps targeting the core network to diminish risk perhaps could be a first step. Next, normalizing the teen to a stable housing environment may reduce the risk of becoming more deeply embedded in social networks that support high-risk behavior. But in the work of being connected to other people, the low-risk teens may help each other or could transfer into the high-risk group over time. But policymakers must remain attentive to the systemic changes can flow from targeting any component of the network. People come and go into each other's lives. Policy must be mindful that the longer the teen is outside of a traditional household, the more human connections will be made with the people that they have the most contact with. Could the teens in periphery have formed cliques that supported less risk taking or conversely supported more risk taking? The teens are tied together by something stronger: love, support, concordance in values, and shared protective as well as risky behaviors (see Feld and Carter 1998, Kadushin 2012). In other words, people live by forming bonds wherever they land. When teens land without a consistent source of shelter, teens huddle for physical and emotional warmth. Once a teen chooses to expend energy to deepen relationships, this act takes time and resources from already deleted stock of navigating instability. It is easier to choose a new focus that is tied in some way to the foci they already share (Feld 1981).

If the policy lumps the core and periphery together, network membership changes over time which could influence the ability of the policy to have the desired effect. Teens that tie together two completely separate networks are called bridges. By theory, the separate networks would not have connected if not for this new bridge. A teen could theoretically build enough prestige and power to convince two divergent groups of homeless teens to join forces (Granovetter 1973). A necessary element to move from the act of convincing into action is conformity. McCulloh (2013) noted that there was a noticeable lack of research that sought to uncover the influence of social network position on conformity. According to McCulloh's (2013) replication of Asch-type conformity tests on two groups of military personnel, actors in the core were less likely to conform and were also more apt to deviate from the social norms without fear of loss of status. The military subjects who were least connected and in the periphery were more likely to conform (McCulloh 2013).

Being on the outside (periphery) of the homeless core appears to protect against HIV risk-taking (Rice et al 2012). The peripherals may be at risk in other ways that may lead to a greater risk of HIV risk taking once the teen is in the core. But there is also a possibility that the teen will become enveloped and become high risk himself. But the possibility illuminated by social network analysis is significant enough to take notice of elements of social embeddedness and social compliance among homeless teens. However, it may be too much to ask of that teen to work to overhaul the collectively held value of higher-risk sexual practices (Long et al 2013). That is where social agencies and interventions must play a sizable role.

In light of the conformity research, the following possible systemic factors may be considered when exploring the structural context of homeless teens' survival networks:

- Every homeless teen is not the same and each with present different social orders.
- Being deeply connected in the homeless culture (core) may be attributable to increased risk for unsafe sexual behaviors.
- Flip the switch on foci among teens that reinforce risky sexual behaviors to foci that are more supportive of positive health choices.
- Targeting low-risk teens on the periphery should account for the higher likelihood to conform to the core values for acceptance.
- While there may be opportunities to use low-risk teens as "bridges" to the high-risk teens, this should only be done with extreme care and oversight. The bridge is more susceptible to falling into the activities of the core and may suffer from burn-out for the heightened sense that change is on that teen's shoulders.
- Watch the movement of teens from the core to periphery (and back again). This movement brings a whole new set of structural realities both for the teen as well as the core and peripheral networks.
- Assess the social purgatory effect on peripheral teens. Target prosocial connections of the periphery which may support a return to a stable living environment.
- Teens in the core, due to their influential status, may be more willing and able to deviate from the prevailing social norms without losing clout with the others.

Social networks are powerful and are often underutilized in uncovering the underlying structure of health policies. The policy work must be held to high standards due to its position in at the forefront of combating ecological gaps and failures. Sometimes the lesson radiates from the personal and societal failing of one HIV-positive homeless teen.

References

Aidala A, Cross JE, Stall R, Harre D, Sumartojo E (2005) Housing status and HIV risk behaviors: implications for prevention and policy. AIDS Behav 9(3):251–265

Allen JP, Porter MR, McFarland FC, Marsh P, McElhaney KB (2005) The two faces of adolescents' success with peers: adolescent popularity, social adaptation, and deviant behavior. Child Dev 76(3):747–760

Asch S (1956) Studies of independence and conformity. A minority of one against a unanimous majority. Psychol Monogr 70(9):1–70

Barnes JA (1972) Social networks. Addison-Wesley Modul Anthropol 26:1–29.

Borgatti S (2005) Facilitating knowledge flows. http://www.socialnetworkanalysis.com/knowledge_sharing.htm. Accessed 8 June 2014

Borgatti S, Everett M (1999) Models of core/periphery structures. Soc Netw 21:375–395

Borgatti S, Halgin D (2011) On network theory. Organ Sci 22(5):1168–1181

Borgatti S, Mehra A, Brass D, Labianca G (2009) Network analysis in the social sciences. Science 323(5916):892–895

Burt R (2004) Structural holes and good ideas. Am J Sociol 110:349–399

Burt R (2005) Brokerage and closure: an introduction to social capital. Clarenden lectures in management studies. Oxford University Press, Oxford

Carpenter D, Esterling K, Lazer D (2003) The strength of strong ties—a model of contact making in policy networks with evidence from U.S. health politics. Ration Soc 15(4):411–440

Cooper A-J (1988) A voice from the South. Oxford University Press, New York

Denning P, DiNenno E (2010) Communities in crisis: is there a generalized HIV epidemic in impoverished urban areas of the United States? Centers for disease control. Presented at: XVIII international AIDS conference, Vienna

Feld S (1981) The focused organization of social ties. Am J Sociol 86(5):1015–1035

Feld S, Carter W (1998) Foci of activities as changing contexts for friendship. In: Adams R, Allan G (eds) Placing friendship in context. Cambridge University Press, Cambridge

Fingerman K, Birditt K (2003) Do age differences in close and problematic family ties reflect the pool of available relatives? J Gerontol B Psychol Sci Soc Sci 58(2):80–87

Granovetter M (1973) The strength of weak ties. Am J Psychol 78(9):1360–1380

Green H, Tucker J, Golinelli D, Wenzel S (2013) Social networks, time homeless, and social support: a study of men on Skid Row. Netw Sci 1(3):305–320

Halley J, Winkler D (2008) Classification of emergence and its relationship to self-organization. Complexity 13:10–15

Hanneman R, Riddle M (2005). Introduction to Social Network Methods. http://faculty.ucr.edu/~hanneman/nettext/. Accessed 20 June 2014

Hawe P, Webster C, Shiell A (2004) A glossary of terms for navigating the field of social network analysis. J Epidemiol Community Health 58:971–975

Kadushin C (2012) Understanding social networks. Oxford, New York

Kidder D, Kidder DP, Wolitski RJ, Campsmith ML, Nakamura GV (2007) Health status, health care use, medication use, and medication adherence among homeless and housed people living with HIV/AIDS. Am J Public Health 97(12):2238–2245

Kilduff M, Tsai W (2003) Social network and organizations. Sage, Los Angeles

Krackhardt D (1992) The strength of strong ties: the importance of philos in organizations. In: Nohria N, Eccles R (eds) Networks and organization. Harvard Business School Press, Boston

Laumann EO, Marsden PV, Prensky D (1983) The boundary specification problem in network analysis. In: Burt S, Minor MJ (eds) Applied network analysis: a methodological introduction. Sage, Los Angeles

Long J, Cunningham F, Braithwaite J (2013). Bridges, brokers and boundary spanners in collaborative networks: a systematic review. BMC Health Serv Res 13:158

McCulloh I (2013) Social conformity in. Netw Conn 33(1):35–42

Meadows D (2008) Thinking in systems: a primer. Chelsea Green, River Junction

Milloy M, Brandon M, Montaner J, Wood E (2012) Housing status and the health of people living with HIV/AIDS. Curr HIV/AIDS Rep 9(4):364–374

Mizuno Y, Purcell D, Borkowski T, Knight K, the SUDIS Team (2003) The life priorities of HIV-seropositive injection drug users: findings from a community-based sample. AIDS Behav 7:395–403

Newman N (2003) The structure and function of complex networks. SIAM Rev 45:167–256

Paranyuskin D (2010) The inclusive exclusivity of the "Dynamique d'Enfer". http://deemeetree.com/current/inclusive-exclusivity/. Accessed 19 Dec 2013

Persitz D (2010) Power and Core-Periphery Networks. SSRN. http://papers.ssrn.com/sol3/papers.cfm?abstract_id=1579634. Accessed on April 9:2014

Rapport A (1953) A marginalized network highlights both the mathematical finiteness as well as socio-political constriction on life chances. Bull Math Biophys 15:523–533

Rice E, Barman-Adhikari A, Milburn N, Monro W (2012) Position-specific HIV risk in a large network of homeless youths. Am J Publ Health 102(1):141–147

Shiell A, Hawe P, Gold L (2008) Complex interventions or complex systems? Implications for health economic evaluation. BMJ 336:1281. doi:http://dx.doi.org/10.1136/bmj.39569.510521.AD

Strogatz S (2001) Exploring complex networks. Nature. http://www.nature.com/nature/journal/v410/n6825/pdf/410268a0.pdf. Accessed 15 March 2014

Trotter R (1999) Friends, relatives and relevant others: conducting ethnographic network studies. in. In: Schensul J, LeCompte M, Trotter R, Cromley E, Singer M (eds) Mapping social networks, spatial data and hidden populations. AltaMira, Lanham

Umberson D, Croscoe R, Reczek C (2010) Social relationships and health behaviors over the life course. Annu Rev Sociol 36:139–157

Valente TW, Hoffman BR, Ritt-Olson A, Lichtman K, Johnson CA (2003) Effects of a social-network method for group assignment strategies on peer-led tobacco prevention programs in schools. Am J Public Health 93(11):1837

Wasserman S, Faust K (1994) Social Network Analysis: Methods and Applications. Cambridge University Press, New York

White D, Houseman M (2002) The navigability of strong ties: small worlds, tie strength, and network topology. Complexity 8(1):72–81

Chapter 4
Communication and Politics in Healthcare

The Policy Puffin Quarrel in Political Colonies

In a biological ecosystem, its structure is composed of its organisms which assure sustainability of its species by means of social relationships among those species. At its more basic level, animal species form groups through attraction with lead to dense clustering (Hemelrijk 2013). When animals cluster, the evolutionary imprint is to use competition to add new members (Hemelrijk 2013). Organisms gravitate and aggregate most often toward the center core and for less time in the periphery (Hemelrijk 2013). According to Hemelrijk (2013), the nature of clustering forms two scenarios:

1. One common cluster populated by all organisms
2. Two clusters formulating a core and periphery

There is a school of thought that intentionality and self-organization are introduced, the animal behavior enters the realm of complex social behavior (Hemelrijk 2013). This is not unlike humans.

A puffin (*Fratercula arctica*) is a North Atlantic seabird, known for nesting as territorial colonies. Colonizing appears to be an exclusive behavior found only in seabird species (Lack 1966). Policymakers join forces and take sides. Policymakers could be described as tufted, highly social puffins, sorted into large cultural, ideological colonies. Orchestration of federal, state, and local policy leadership involves decision-making across players between agencies and within agencies before the outside is brought in. As "puffins" of different jurisdictions and political proclivities, policymakers of all stripes serve a distinct and noble purpose of changing the social course in health care. Policymakers strategize, either in an environment of accord or, more often, in disagreement with other stakeholder puffins. Some win through social advantage due to real and perceived hierarchical advantages. The policy puffins' social organization is the epitome of homophily better known as "birds of a feather" (see Lazarsfeld and Merton 1954; Marsden 1988). According to Feld and Carter (1998), homophily is socially accepted commonality of group membership such as that of the legion of organized policymakers. Puffins find security

© Springer International Publishing Switzerland 2015
M. Battle-Fisher, *Application of Systems Thinking to Health Policy & Public Health Ethics*, SpringerBriefs in Public Health, DOI 10.1007/978-3-319-12203-8_4

in colonies of other like-minded or like-assigned policy puffins. Since puffins are a biologically homogeneous colony, how might differences be made in understanding dominance hierarchy? Victories within animal species appear to be self-reinforced by past wins. As the winning is based on past social dominance, the prestige of such wins is not diminshed by losses to low-ranking group members (Hemelrijk 2013).

When a puffin is approached by an intruder (counterargument), the puffin hunkers in. It is apparent that people disagree and that disagreement may happen across party lines. Puffins in this phenomenon may be framed in terms of homophily based on attitude. The puffin, flashing its snow white chest plumage, spreads its wings and kicks. The instinctive strength in position by the puffin is accompanied with disciplined wrestling to gain leverage. Just because people choose to be around each other that does not mean that they will believe and act like the others. Peer effect indicates that there is a social force that supports people believing and acting like each other. Policy puffins, for the sake of this argument, are sorting themselves to be around policymakers like them. There is less perceived trepidation in homogeneity. The sharing of values in part tells a story about the nature of political-based stalemate.

Allies are needed to develop a united bastion to fend off threats from political adversaries (Smith et al. 2014). In fact, being under political siege may not be a bad thing for the influencer in power. When under attack, the person with the most political power derives prestige based on the congregation of allies with no other recourse for protection (Smith et al. 2014).

> Evidence that one party regards as devastating to a second party's argument, the second may dismiss as innocuous or irrelevant. (Schön and Rein 1994)

Goel et al. (2010) described attitude homophily, that is analogous to being in attitude agreement. There is often a misconception afoot within clusters of allies among that all are in agreement on an issue based solely on allegiance (Goel et al. 2010). A false consensus occurs when an overestimated leap is made regarding the nature of a person's private political stance made in the absence of information to the contrary (Krueger and Clement 1994). For instance, misguided assumptions of allegiance can occur where inference is made that all moderates of a particular political party would affirm the same spirit of positions to the same degree and in every case. Holding a dissenting position (or attitude) may not materialize into a dissenting vote that counters the party's positions (see Goel et al. 2010).

Policymakers take cues from social and political energies that are often understood as the successes and failures realized from the vantage afforded as the agent of governance. But this may offer an incomplete or even incorrect explanation of how policy works once the policy regulation is released into the public. Colander and Kupers (2014) in their formative work call for the use of complexity theory to elevate market economics to public economics. The liberal top-down blanket of governmental control and the conservative explanation of free markets are devoid of any appropriate responses to the fact that government is but one of "endogeneously evolved control mechanism" (Colander and Kupers 2014). Any sophistication that recognizes that top/down mechanisms and markets are symbiotic is lost within the puffin fight (Colander and Kupers 2014). Strict adherence of political homophily at all costs breeds an environment where attribution errors and snuffing of novel information into the system can harm the public policy. The chasm between political networks is more than ideological. The fissure between puffins due to homophily ignores complexity in public policy to its grave detriment.

The Flow of Ideas within Networks: Focus on Interprofessional Communication

Granovetter (1978) presented the idea the *threshold model of collective behavior*. In accordance to this threshold model, binary decisions such as pro tobacco control and con tobacco control is based on a threshold of others' participation in the deliberation. Accounting for the element of direct communication, Valente (1996) also added to the network lexicon network thresholds, which are measured in terms of exposure to *direct communication ties* but not as a threshold measured for the whole social system. This threshold shows the point where a node is more convinced to adopt an idea. This would be important to know if a person may be ethically malleable meaning that ethical stances are influenced by others and are not resistant to change (Battle-Fisher 2010). Valente (1996) found that opinion leaders have lower network thresholds; they need the least convincing to innovate. However, these opinion leaders must find a way to influence a more resistant node in the network. These people would be those later adopters that are hard to reach and crack. But often in health, we do not have enough buy-in of the public. If that public is bounded by smaller, more approachable networks, we have a chance. If that influential network bridges to other networks to diffuse an ethic, then we have lift-off. Now how that ethic is being diffused is another story.

> We have a natural tendency to romanticize breakthrough innovations, imagining momentous ideas transcending their surroundings … But ideas are works of bricolage … We take the ideas we've inherited or that we've stumbled across, and we jigger them together into some new shape. (Johnson 2010)

Physical closeness in social networks is known as *propinquity*. A requirement of propinquity is being in the same place at the same time (Kadushin 2012). People near each other form bonds and possess similar social characteristics, such as co-workers in the same department division or smokers on the same smoke break every day. Some find trouble making or even desiring to make new ties to innovate or gain new information. *Bond forming* differs from propinquity. Sharing a common characteristic, such as an idea, norm, ethic or social attribute, is best framed as an issue of homophily (Kadushin 2012). Therefore adoption of an idea must go beyond shared space and serendipity. Granovetter (1973) found that, in general, it was the existence of weak ties in networks that had the more positive influence on change as there was no competition for scarce resources. Weak ties have the least invested. However, there are policy experts that do invest but step back as well. Brokering so extensively can get exhausting even for the most dedicated supporter. Sources of good ideas are expected to be bridges between networks (Burt 2004).

Political Expediency: Influencing Ideas

The importance lies in the layering of the overarching political landscape to the connectivity within these networks. Individuals will be most influenced by their in-group when saying yea or nay to an idea (Mackie 1986). Inherent to the ability to influence must be an inclusion of the factor how the information (idea) actually flows within a network. Simply, word of mouth communication involves the flow of information between an information sharer and a receiver. To this requirement of word of mouth, Frenzen and Nakamoto (1993) noted two elements of the theory of information flow: the decision and the structure. Frenzen and Nakamoto (1993) noted that both tie strength and opportunity cost demonstrated as pro-idea versus con-idea would affect information sharing. Fundamentally, during word of mouth, words could be heard but if the determination of worth lay squarely with the receiver.

In illustration, of all the possible ties in the Department of Health and Human Services (DHHS) agencies involved in tobacco control, Leischow et al. (2010) found that only 16f% were used. Size of the network is a cumulative count of the number of nodes in the network graph. But the most central bears the burden of over-reliance on a few agencies to conduct enough communication of behalf of the leadership network at large to not compromise DHHS' work in tobacco control. This is also true for the structural integrity of networks. These connections need to be physically as well as emotionally accessible. In the figure, a tie is shown if two agencies communicated at least once quarterly. High centrality, which is a high-level power and reliance in the hand of a node, is a heightened factor in network failure. This translates to power concentrated in the hands of very connected nodes and places the health of the network on the backs of those nodes.

The systemic tale from the DHHS research is in the protection of the integrity of the network. For the sake of governance, all the agencies by principle were to be involved in varying degrees in regulating tobacco. In terms of network connectivity, according to Miller and Page (2007), the "deep" quality of the complexity would mean that the parts of a sum of the networks will have structural repercussions on the health of the entire system (the sum). However, actor and tie failures are the main culprits to its overall stability. A breakdown in a major throughfare for information or the loss of a pivotal agency's involvement could spell trouble. If the agencies that are overly relied upon, fail to continue to act as a catalyst maintaining the system, the systemic effects could be immense. This is particularly true if fail-safes have not been put in place by policy to fill these gaps or create equivalent detours to assure the network works (Figs. 4.1 and 4.2).

Case for Your Consideration: US Tobacco Control Policy and Regulatory Networks

Using The United States' The Family Smoking and Prevention Act (Public Law 111–31) as an example, systems inherent to anti-tobacco policy and how unexpected factors emerge once such a divisive policy is enacted. Passed during President

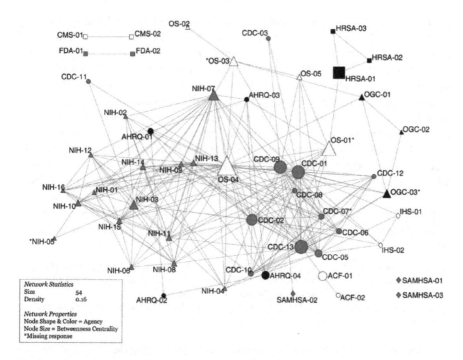

Fig. 4.1 Network of communication contacts among Department of Health and Human Services tobacco control leadership (network tie indicates contact at least once a quarter). *ACF* Administration for Children and Families, *AHRQ* Agency for Healthcare Research and Quality, *CDC* Centers for Disease Control and Prevention, *CMS* Centers for Medicare and Medicaid Services, *FDA* Food and Drug Administration, *HRSA* Health Resources and Services Administration, *IHS* Indian Health Service, *NIH* National Institutes of Health, *OGC* Office of the General Counsel, *OS* Office of the Secretary, *SAMHSA* Substance Abuse and Mental Health Services Administration. (Reprinted from Leischow et al. 2010, with permission from Taylor & Francis Ltd)

Agency	n	Lack of time (%)	Organizational structure/bureaucracy (%)	Interagency politics (%)	Unable to identify appropriate collaborator (%)	Incompatible agency goals or strategies (%)	Benefits of collaborating do not outweigh costs (%)	Past experiences (%)
CDC	11	82	73	36	36	27	27	9
NIH	15	87	47	47	33	20	20	20
OS	3	33	100	0	33	33	0	0
All others	20	80	45	15	20	25	20	10
Total	49	80	55	29	29	25	20	12

Note. CDC = Centers for Disease Control and Prevention; NIH = National Institutes of Health; OS = Office of the Secretary.

Fig. 4.2 Reported barriers to collaboration with other agencies in the DHSS network. (Reprinted from Leischow et al. 2010, with permission from Taylor & Francis Ltd)

Barack Obama's first administration, The Family Smoking and Prevention Act (Public Law 111–31) as an addition to new Chapter IX to the Food, Drug, and Cosmetic Act became the most sweeping regulatory support in combating a leading cause of preventable death in the USA that have been scientifically linked to tobacco use. The Family Smoking and Prevention Act did not take away access to tobacco to consumers of age. Tobacco is sold legally, under the continued regulatory oversight from the Federal Food and Drug Administration. Newly 'strengthened' regulatory powers supplied by this act works to ensure the "safest" possible tobacco products sold to US consumers using the most ethical forms of marketing to the public. Safe is certainly used loosely here. The sale of tobacco as ethical is another question. But tobacco is legal to use for those of age to use the product. What is allowed from a regulatory standpoint by this Act is targeting sales, marketing and distribution channels. Upon maximization, the virtuous goal is to optimize the desired benefit of tobacco control policy, thereby saving lives and improving quality of life from morbidity and mortality related to tobacco use.

> "This legislation will not ban all tobacco products, and it will allow adults to make their own choices ... We know that even with the passage of this legislation, our work to protect our children and improve the public's health is not complete."
> —President Barack Obama during the signing of the Family Smoking Prevention and Tobacco Control Act (2009)
>
> Deliberate or orderly steps are not an accurate portrayal of how the policy process actually works. Policy making is, instead, a complexly interactive process without beginning or end (Lindblom and Woodhouse 1993).

President Obama's statement illustrates the depth and limitations of the powers of policy and portends systemic mess that could possibly ensure even under the most noble of intention. From his statement, we can take away that the following issues are central to the prescript of the law.

1. Increasing regulation of a legal product, in this case, tobacco.
2. Blocking tobacco advertisements from children within a specified distance of schools and parks.
3. Making flavored tobacco distribution illegal. Menthol is excluded from this ban.
4. All harmful chemicals must be fully disclosed to the smoker along with graphic fear appeals on the labeling.
5. Public Safety overrides autonomous actions of the consumer. Some elements of the society, such as children, require higher ethical safeguards which require the power of policy behind it.

Is an advertisement on the way home from the park that falls outside the banned quadrant acceptable? By the letter of the policy, it is allowed. Is it ethical at the point of exposure to the ad or to the mere possibility of exposure to the marketing?

Cigarette smokers have been found to have very high levels of brand loyalty (see Dawes 2012). The smoker knows the brand, down to the flavor, even recalling the specifications of the box that they want. At times, perhaps another perceived equivalent cigarette will do just this once and buy a comparable brand if the requested one is not available (Dawes 2012). For others, not having their brand is a game breaker. Dawes (2012) in his analysis of US cigarette consumer panel data uncovered the strong power of perception in cigarette branding.

1. A smoker that purchases high end cigarettes most likely will not lower the expectation of the quality perception by buying a generic brand.
2. If there is a female aesthetic on the box, men will not bite at all even when the call of nicotine gnaws. They will look for a cowboy or something testosterone driven in marketing appearance.
3. Price point does matter.

If you want overwhelming consumer loyalty, market to a smoker. (Dawes 2012)

References

Battle-Fisher M (2010) Organ donation ethics: are donors autonomous within collective networks? Online J Health Ethics 6(2). http://aquila.usm.edu/ojhe/vol6/iss2/6. Accessed 14 April 2014

Burt R (2004) Structural holes and good ideas. Am J Sociol 110:349–399

Colander D, Kupers R (2014) Complexity and the art of public policy-solving society's problems from the bottom-up. Princeton: Princeton University Press

Dawes J (2012). Cigarette brand loyalty and purchase patterns- an examination using U.S. consumer panel data. Available at http://ssrn.com/abstract=21226951. Accessed on 14 April 2014

Feigenbaum M J (1983) Universal behavior in nonlinear systems. Physica 7:16–39

Feld S, Carter W (1998) Foci of activities as changing contexts for friendship. In: Adams R, Allan G (eds) Placing friendship in context. Cambridge University Press, Cambridge

Frenzen J, Nakamoto K (1993). Structure, cooperation and the flow of market information. J Consum Res 20:160–175

Goel S, Mason W, Watts D (2010) Real and perceived attitude agreement in social networks. J Pers Soc Psychol 99(4):611–621

Granovetter M (1973). The strength of weak ties. Am J Psychol 78(9):1360–1380

Granovetter M (1978) Threshold models of collective behavior. Am J Sociol 83(6):1420–1443

Hemelrijk C (2013) Simulating complexity of animal social behaviour. In: Edmonds B, Meyer R (eds) Simulating social complexity—a handbook. Springer, New York

Johnson S (2010) Where good ideas come from. Riverhead-Penguin, New York

Kadushin C (2012) Understanding social networks. Oxford University Press, Oxford

Krueger J, Clement RW (1994) The truly false consensus effect: an ineradicable and egocentric bias in social perception. J Pers Soc Psychol 67(4):596–610

Lack D (1966) Population studies of birds. Oxford University Press, Oxford

Lazarsfeld P, Merton RK (1954) Friendship as a social process: a substantive and methodological analysis. In: Berger M, Abel T, Page C (eds) Freedom and control in modern society. Van Nostrand, New York

Leischow S, Luke D, Mueller N, Harris J, Ponder P et al (2010) Mapping U.S. government tobacco control leadership: networked for success? Nicotine Tob Res 12(9):888–894

Lindblom C, Woodhouse E (1993) The Policy-Making Process. Prentice-Hall, Englewood Cliffs, NJ

Mackie D (1986) Social identification effects in group polarization. J Pers Soc Psychol 50:720–728

Marsden PV (1988) Homogeneity in confiding relations. Soc Netw 10:57–76

Meadows D (2008) Thinking in systems: a primer. Chelsea Green, White River Junction

Miller J, Page S (2007) Complex adaptive systems: an introduction to computational models of social life (Princeton Studies in Complexity). Princeton University Press, Princeton

Schön DA, Rein M (1994) Frame reflection. Basic Books, New York

Smith K, Christakis N (2008). Social networks and health. Annu Rev Sociol 34:405–429

Smith J, Halgin D, Kidwell V, Labianca G, Brass D, Borgatti SP (2014) Power in politically charged networks. Soc Netw 36:162–176

Valente TW (1996) Social network thresholds in the diffusion of innovations. Soc Netw 18:69–89

Wasserman S, Faust K (1994) Social network analysis: methods and applications. Cambridge University Press, Cambridge

Chapter 5
Health Systems and Policymaking as the *"Price Is Right"*

The Effect of Game Theory on Systems

Game theory uses math and probabilities to predict future results. By definition, game theory deals with the rules of strategy that are in some fashion dependent on external influences in a competitive environment. An example of how people grapple with social complexity could be the popular US game show, The Price is Right. The error in the guess is not revealed in the Cliff Hanger game until after the cardboard yodeler has moved up the mountain illustrating the deviation of the guess from the correct answer. Factors present in the decisions made when playing a game include:

1. The initial mechanism used by the contestant to come to the guess
2. Any dynamic alterations made to subsequent guesses
3. The distance away from the actual price (which is shown in the steps taken by the figure)
4. The distance left from the goal

Berk et al. (1996) wrote a paper, testing whether rational decision theory exists in an environment of substantial economic incentive. An important caveat is that the contestant in the game show must believe that there is a chance of beating the system. As elemental to the games of chances, the contestant had incomplete knowledge to make a successful rational choice (Berk et al. 1996). Imagine the dismay when the contestant is surprised how far away they are from the correct guess. The contestant would continue to bid.

> Hodl-oh-ooh-dee
> *Hodl-ay-ee-dee-*
> *Hodl———-ay–ee-dee-yi—ho.*

The show is based on the fascination of the possibility of success, not its realization of success. Policy is graded by securing success. But what is unique to *The Price is Right* is that the contestant often calls upon the help of the screaming audience for help in making future guesses in real time, particularly if he or she failed on a

© Springer International Publishing Switzerland 2015
M. Battle-Fisher, *Application of Systems Thinking to Health Policy & Public Health Ethics*, SpringerBriefs in Public Health, DOI 10.1007/978-3-319-12203-8_5

previous attempt. In the end, there is a system to work through which may in this case be futile.

External Influences on Decision-Making

Miller and Page (2007) explored the standing ovation problem that dealt with how peer effect can modify behavior. In the Price is Right example, the standing ovation model is concerned with the public displays of social compliance to a group. In this case, the compliances can be demonstrated by standing up in applause or matching verbal expression to match those displayed around them. According to this model, agents are in pressure to make sure that our actions match others around them. But there are choices:

- Some people in the audience are early birds and they stand and clap first.
- Others may choose to not stand or clap at all regardless of the social pressure around them.
- Others may look around either through personal judgment of the performance or through mimicking the behavior of others around them. In the end, each joins in the ovation.

The standing ovation problem is about the power of peer effect.

The *standing ovation problem* presented by Miller and Page (2007) supported the idea of peer effect strongly encouraging the act of applauding in public in accordance to the actions of the group. Individuals have been found to more likely copy an action if there is more pressure to comply with the peer effect. In general, the lower personal threshold to trigger an action (applauding) increased the probability of applauding (Miller and Page 2004, 2007). Intimacy to people in the crowd also mattered. If there is a recognized dyad (such as two people with a relationship in the same public space and time), there is a higher recognized level of peer influence. A person is more likely to match actions with his or her companion in the dyad (Miller and Page 2007). The people in the front and well as celebrities matter to collective behavior (Miller and Page 2007). All the while, if the threshold tip is easier to reach in a group, more may be encouraged and socially protected to stand (Miller and Page 2007). Individuals in the audience are able to enjoy a social protection in a social display if the same action reflects that of the group.

Collective Action

Miller and Page's (2007) rule-based assertions on peer influence differ slightly from that of Mark Granovetter's *theory of collective action* (see Granovetter 1978). In appraising quality in collective action, Granovetter (1978) warned that motives in collective action are not mirrors into the soul of agreed upon norms. When all of the contestants are yelling, there is no pow-wow incoming to a value-based consensus. There is no time or no need to come to a meeting of the minds while Bob Barker prodded for a fast response to the game. There is a commercial promotional break coming up in ten. The audience plays its role as individual agents exerting some influence on the state of affairs. Also, according to Granovetter (1978), collective action is most likely to occur with a lower threshold to act on a binary decision to do it or not.

> Every physical quantity, every part of it, derives its ultimate significance from bits, binary yes-or-no indications. (Wheeler 1989)

Rules imply people are devoid of cognition or autonomous ethical struggles. Here lies the conundrum. But there are no ironclad rules in what agents assert as their decisions. Political decisions are not made within civic squares. Rationality is extremely restricted to achieve. But rule-based models offer choices devoid of the messiness of cognition. A trigger is a trigger. There are no purposeful agents to contend with. The rules have been created to mimic reality as much as possible. It is important to discern if the policy targets actions that have no wider consequence to others or if the driver depends on the actions of other to comply. But in order to understand if people act based on sorting (homophily) or peer effect (collective behavior), the question must be answered if there is only movement/action (homophily) or there is a change afoot supporting that action (peer effect). Is the policy seeking to gain surface insight of policy outcomes or understand the underlying cascading mechanisms of change propagated by that policy? Rules offer a foundation but are not the end of the story.

Example of Living Organ Donation in the USA

For instance, discussions of living organ donation often occur during times of eminent duress of the eminent need for an organ. Networks may cross based on purpose or be marked by isolation. As social beings, negotiation of social agreement often requires personal engagement with people we trust and share strong attachments. If an ethic develops at a larger level, how might success of a positive donation ethic is accomplished if consensus may be made as a collective of individuals that they know? Likewise, how much does one person hold in influencing the ethical beliefs of others around them if the overall moral position contradicts the larger system network level?

Organdonor.gov, in its "Get Started" tips for declaring donation, intentions the point to familiarize your family with your decision. This assumes that one's ethic aligns with the "decision" rendered to others or that there will be a "collective change of heart." As Fox (2010) notes that a less-charged setting would be appropriate for discussing postmortem donation, what of the charged reality that there is a collective ethic that may be working against donation? People in a network by nature are emotionally invested. It would be best to approach this health discussion as it is emotive and difficult to navigate in all circumstances.

This talk enters the process at delayed juncture. They may very well have been socially influenced by the very advocates that were asked to support during the big disclosure. Again, explore the nature of the networks and collective influence. Donors are "potential" even if their bioethical position is contrary (Fox 2010). "It ain't over till it's over" (unless the patient's body fails or everyone in the network rejects donation). It is the tricky part. It may not be fully apparent just how much manipulation and change in a system is required to reach a goal (see Meadows 1999). It would be premature to guess just how many prodonation individuals it would take to make the network tip positively to result in a living organ donation. Taken from another perspective, what is the tipping point that outflows, resulting in losing support? Case in point, human agency and the ability to change one's mind give an ethical malleability that makes changing ethical positions possible and gives clinicians a glimmer of hope for a successful conversion to organ donation.

> Typically, the expert knowledge of the people who actually operate the system is required to structure and parameterize a useful model (Ford and Sterman 1997).
>
> We are unaware of the majority of the feedback effects of our actions. Instead, we see most of our experience as a kind of weather: something that happens to us but over which we have no control (Sterman 2002).

Divergence and Subjectivity—Is it Winning or Losing?

> Perhaps, the real power of stories lies in their reflection of ideas and values … Much of the policy process involves debates about values [ethics] masquerading as debates about facts and data. (McDonough 2001)

According to Schumacher (1977), a divergent solution is one comprised of many intertwined parts that require a hardy fortitude to tackle. Chapman (2004) described two similar types of problems taking to account the particulars of policy. First, Chapman borrowed from Ackoff (1974) stressing the difference between "difficulties" and "messes." Difficulties are convergent with a finite allocation of time and resources to the targeted problem. On the other hand, the "messes" are more elusive in forming consensus and the ideal solution may not materialize (Ackoff 1974; see Chapman 2004). "Messes" such as healthcare policies enjoy the ability to frame and affect the public based on the present knowledge, resources, politics, and social climates in which the policy was enacted. As long as differing positions abound around a health policy issue, a mess of innate nonlinearity must be accounted for (Chapman 2004).

> The habitus—embodied history, internalized as second nature and so forgotten as history is the active presence of the whole past of which it is the product. (Bourdieu 1993)
>
> Complexity theory is unable to address "issues of subjectivity," meaning, the limitations of language, and the essentially interpenetrative and transformative character of human experience (Chia 1998).

Aaron Riley, a longtime social advocate, offered his thoughts on the roles of community stakeholders in affecting policy. Currently, Mr. Riley is the CEO and Founder of New Leaf Columbus in Columbus, Ohio. The mission of New Leaf Columbus (http://newleafcolumbus.ning.com/) is to strengthen LGBTQ (Lesbian, Gay, Bisexual, transgender, and Questioning) communities of color through dialogue and advocacy. With advocates, often peering from the outside into the political process, there is a tension between being heard and affecting observable social change. According to Mr. Riley, "far too many policies lacked both the art [reflexivity] and science" that in turn fundamentally flaw policymaking. He added that advocates and policy makers often find themselves facing a reality of reliance on business as usual that breeds competing priorities and competing factions with little gained for any party. Advocates are aware of the limitation of resources, which in his opinion, "in itself is a powerful generator (to push) policy."

> It is a quagmire for sure; however, in my opinion, it is much worse to do nothing. So do extraordinary circumstances call for extraordinary problem solving? Perhaps, it sometimes just calls on us to learn from what we have tried that did not work in order to turn us in a new or different direction. After all, as imperfect beings in an imperfect world that can only create imperfect policy, the best we can hope for is the evolution of our imperfection. Aaron Riley, interviewed by the author (2014).

Reflexivity is often flatly ignored when examining social systems. Greg Fisher (2012) offered what he called the Law of Ostrichs (Fisher 2012). Contrary to the popular belief, ostriches do not place their heads in the sand at all. In defense of a predator's advance, ostriches in fact run in an attempt to coax the enemy to follow them, so that they (ostriches) can protect their eggs. In terms of its applicability to policymaking, comfort is often taken over the "awkward truth" of what is really happening in a social system (Fisher 2012). The Law of Ostrich's applies "when a comforting yet inaccurate narrative is believed ahead of an awkward truth." (Fisher 2012). Meadows (2008) wrote that stakeholders hold fast to their own bounded rationality of the policy issue. What results is a clash of the rationalities based on the decisions toward acknowledging reflexivity in the policymaking process. The system becomes policy resistance, typified by the "intensification of anyone' effort leads to the intensification of everyone else's" (Meadows 2008). The best approach to abolishing policy resistance is to get rid of the bad policies (Meadows 2008). Some may argue that politics makes this requirement of outside engagement less than realistic.

References

Ackoff RL (1974) Redesigning the future: a systems approach to societal problems. Wiley-Interscience, New York

Berk J, Hughson E, Vandezande K (1996) The price is right, but are the bids? An investigation of Rational Decision Theory. Am Econ Rev 86(4):954–970

Bourdieu P (1993) Structures, habitus, practices. In: Lemert C (ed) Social theory: the multicultural and classic readings. Westview, Boulder

Chapman J (2004) System Failure. Demos. http://www.demos.co.uk/files/systemfailure2.pdf. Accessed on 15 Jan 2014

Chia R (1998) From complexity science to complex thinking: organization as simple location. Organization 5(3):341–369

Fisher G (2012) Reflexivity and Narrative. http://www.synthesisips.net/blog/reflexivity-and-narratives/. Accessed 8 June 2014

Ford D, Sterman JD (1997) Expert knowledge elicitation to improve mental and formal models. Syst Dyn Rev 14:309–340

Fox M (2010) Organ donation should be a part of health discussions. American Medical News. http://www.ama-assn.org/amednews/2010/05/31/prca0531.htm. Accessed 14 April 2014.

Granovetter M (1978) Threshold models of collective behavior. Am J Sociol 83(6):1420–1443

McDonough J (2001) Using and misusing ancedote in policy making. Health Aff (Millwood) 20(1):207–212

Meadows D (1999) Leverage points: places to intervene in a system. The Sustainability Institute, Hartland

Meadows D (2008) Thinking in systems: a primer. Chelsea Green, White River Junction

Miller J, Page S (2004) The standing ovation problem. Complexity 9(5):8–16

Miller J, Page S (2007) Complex adaptive systems: an introduction to computational models of social life (Princeton Studies in Complexity). Princeton University Press, Princeton

Schumacher E (1977) A guide for the perplexed. Harper & Row, New York

Sterman J (2002) All models are wrong: reflections on becoming a systems scientist. Syst Dynam Rev 18:501–531

Wheeler JA (1989) Information, physics, quantum: the search for links. Proceedings III International Symposium on Foundations of Quantum Mechanics-Tokyo, pp 354–368

Chapter 6
Ethical and Systematic Approaches to Health Policy

Ethics and Health Policy

As long as a person exists in the presence of other carbon life forms, ethics will be batted around. Ethical malleability acknowledges that a person may "change" his ethical stance to suit his present belief system (Battle-Fisher 2010). The private element of self-reliance perculates under the same ecological conditions as the public concern for others (see Colander and Kupers 2014). Brass et al. (1998) introduced the idea of using social network analysis to understand ethical negotiation.

> The [effective] government does not impose norms [ethics] or even force individuals to self-regulate. Instead it attempts to encourage the development of an econstructure that encourages self-reliance and concern for others. (Colander and Kupers 2014)

There could be a few ramifications in terms of demonstrating a united ethical front of the collective for crossing a moral boundary. Social influence, not to be confused with coercion, may act as a morality azimuth. Does the affirming of bioethics require relevance, novelty, or something else altogether? Susceptibility to opinion leaders' beliefs may be due to resultant centrality within a particular network. There are true power brokers who may sway other enough to influence a choice in ethical decisions. Does this counter the ideal of autonomy? No, his malleability serves as a reality check to often unobtainable true autonomy. While people remain social creatures, autonomy will be a striving toward a pinnacle of self-realization. We operate along that journey toward this heightened state of actualization; therefore, we are ethically susceptible to malleability. Might an ethic be "innovative" with the ability to diffuse across a network? Could a person change his mind and reverse the "innovation" or is it just a new feedback loop into the cycle of innovation with no consequence?

For instance, if a person has a high measure of trust plus a high level of closeness, which makes it easier to find others to influence, imagine the ability to influence ethics under this scenario. Is there really no consequence for not accepting a diffused ethical position posed by opinion leaders? An ethic can be changed until it is acted upon. Even after it is acted upon, a person may regret that moral decision then revert and have to deal with the collateral damage of that choice. An ethic is

© Springer International Publishing Switzerland 2015 57
M. Battle-Fisher, *Application of Systems Thinking to Health Policy & Public Health Ethics*, SpringerBriefs in Public Health, DOI 10.1007/978-3-319-12203-8_6

not the action but a precondition to action. But the public ultimately must pay for its actions so the public indirectly pay for personally held ethical positions.

As rationality serves a purpose in framing situations, people are owners of active frontal lobes. Granovetter (1985) introduced the fundamental idea of the consequence of embeddedness on social relationships. The private self is folded into a public one. A person does not act autonomously when under the influence and prying eyes of one's network (public). "Concrete personal relations and structures (or networks) of such relations" guide members of a system (Granovetter 1985). While struggling to maintain the sanctity of personal agency, people are influenced by others in a position of social influence. An ethic may never necessarily be displayed as a discernible action or even need to be articulated. Must an ethic be communicated to count? Is it something else when discussed by morphing into a value or norm?

> Only if (punitive) consequences can sometimes override such factors in determining morally appropriate actions do the cases illustrate moral mistakes. (Adler 2005)

> Morally conscientious government actors must ultimately settle on a moral theory (or a probability distribution across theories) and choose. (Adler 2005)

As the public engages with its environment , "good" or "moral" changes are tied to the overall state of a public health issue. Law provides government institutions the authority to act in enforcement, thus enabling healthy personal choices. Law cannot dictate, though it may influence the nature of personally acted-upon ethics. There is no direct recourse of the legal authority to the nature of a held ethic insofar that the actions connected to an ethic do not cause an obvious violation of legal standard. Until there is a legal violation, I argue that people have the right to change ethics and often act upon that right. When these decisions happen, the state of the moral system is thus affected.

Interplay of Politics and Ethics

Politics and ethical decisions comingle insomuch as the explicit rules come under constant scrutiny by a public who themselves stay "malleable" based on their own exogeneous factors that pull at them as well (Battle-Fisher 2010). The political process explicitly works in an environment of competing interests. The interests do not solely compete based on ideological group-based differences at the macro level. This is the tenuous environment in which policy functions, one that buoys rules existing under the constraints of political infeasibility and the ever-present possibility of public fatigue toward regulatory oversight. For policymakers, the scuffle becomes one based on leveraging expertise and influence to make marks on policy. After the scuffle subsides, the policy puffins are proud puffins with a common goal but displaying allegiance with sameness. Policymakers navigate a primal system of conquering to advantage or at the least, staying relevant in the debate.

There are drawbridges for moral hazards by influential actors which control the flow of information (Frenzen and Nakamoto 1993). But this flow is not done in a vacuum. In accordance to an information cascade we obtain private information from the "environment" (Bikchandani et al. 2002). If information is not revealed into the environment that is then encoded into discourse, then there would be no matter onto which consensus is made. There will be no new information infused into the network with the chance to affect the tide of decision making. An essential element to ethics is that there must be some data in the environment that was labeled as required to make a determination of principle. Principled action is considered such based on the social agreement that regulation of behavior is inherent to being a member of the public. The higher the moral risk of the idea, the greater the liquid network would need to work at building strong cohesion to offset bailing out of supporting the idea. In other words, will communicating this idea come back to haunt me later due to its moral risk? Is this decision worth the political fallout? Is time over a coffee break long enough to negate the stumbling block of moral risk?

The boundary as a mode of reorganization may work to keep the most involved actors in much like a semipermeable membrane of a cell. In passive cellular transport, the simplest type, diffusion, involves the movement of "molecules" from higher to lower concentration to maintain equilibrium. As people come and go across social boundaries, equilibrium is structural cohesion. As people break ties, the balance may be thought of as maintaining the cohesion of the group as optimally as socially and politically possible. But honestly how difficult is it for people to come together and stay together with a common (and stressful) cause? When agents are brought together by cooperation, structural cohesion is then framed as relational solidarity (Kilduff and Tsai 2003). How can policy serve to support the development and maintenance of ties and role adoption? Ethical malleability may be influenced by positive or negative social energies. People are there for each other; then they are not. A person leaving a network may theoretically serve to consolidate more intensely the resources toward the center, while simultaneously depleting the net resources at their disposal.

These ideas of the core to the periphery are not new metaphorically. Irish poet and dramatist William Butler Yeats composed in the aftermath of World War I in "The Second Coming" as an examination of the instability of the social and moral centers:

> The falcon cannot hear the falconer:
> Things fall apart:
> The centre cannot hold

Yeats metaphorically described the center and periphery as dynamic (Deane 1995). Most importantly, if society is a whole, its unity and coherence—even its very identity—is dependent on the integrity of the center. For if the center is removed, the peripheral parts will no longer join together to form a whole by means of structural cohesion. Moreover, the continued unity of the whole depends on the strength of the center, that is, its ability to hold the periphery in place (Deane 1995). How might one take Deane (1995) out of the world of "ordinary language" of metaphor of the center as influencers in tobacco control and into the universe of mathematical complexity? While everyone may not have the resources to act upon their inclination, there is will always be some level of disharmony. In other words, people do not follow set rules very well. Sometimes the rules are not made by public required to follow them.

References

Adler M (2005) Cognitivism, controversy and moral heuristics. Behav Brain Sci 28:542–543

Battle-Fisher M (2010) Organ donation ethics: are donors autonomous within collective networks? Online J Health Ethics 6(2). http://aquila.usm.edu/cgi/viewcontent.cgi?article=1085&context= ojhe. Accessed 14 April 2014

Bikchandani S, Hirshleifer D, Welsh I (1992) A theory of fads, fashion, custom, and cultural change as informational cascades. J Polit Econ 100:992–1026

Brass D, Butterfield K, Skaggs B (1998) Relationships and unethical behavior: a social network perspective. Acad Manag Rev 23(2):14–31

Colander D, Kupers R (2014) Complexity and the art of public policy-solving society's problems from the bottom-up. Princeton University Press, Princeton

Deane P (1995) Metaphors of center and periphery in Yeats' the second coming. J Pragmat 24:627–642

Frenzen J, Nakamoto K (1993) Structure, cooperation and the flow of market information. J Consum Res 20:360–375

Granovetter M (1985) Economic action and social structure: the problem of embeddedness. Am J Sociol 91:481–510

Kilduff M, Tsai W (2003) Social network and organizations. Sage Publications, Los Angeles

Chapter 7
Health Disparities in Public Health

Organ Donation in a Complex System

Berlinger (2009) spoke eloquently of the dogged issue of the perpetuation in flawed reasoned (in)action within complex systems. If complex systems defy description, the personal social network of the potential donors must be a part of the discussion of supporting living organ donation (Berlinger 2010). Thank goodness that no man or woman is an island. Other people (and their valuable organs) are needed in consort with organ donation. The norms of living organ donation may be understood by mapping a network of close confidants (known as nodes or actors) that are linked by a particular circumstance. The central character or ego (ironically named so) is not alone with his or her thoughts. The ego is connected to others in a network whether large or small in size. The self-centeredness of an "ego's" ethical decision becomes complicated by its embeddedness in a network of concerned others. To embed socially leads to an overlap of roles of private lives.

Health disparity of organ donation is often explained with statistics. This is only a part of the narrative: enumeration as foreign to the ego as the connectedness to the entire population of end-stage renal disease (ESRD) patients that only emotionally reach as far as those they know and care about. Each family is aware of the finiteness of opportunity cost for their loved one; a kidney transplanted to another without a donor in replacement lengthens the odds of a miracle. Most living donations come from biological-related donors. This will be the network. There is a layering of judgment of morality. An event that may begin in earnest as an autonomous act is no longer so. One must account for the *heteronomy*, or difference in values that may be originated and perpetuated by the network.

Earlier versions of this chapter appeared as Battle-Fisher M (2010) Organ donation ethics: are donors autonomous within collective networks? Online J Health Ethics 6(2). http://aquila.usm.edu/ojhe/vol6/iss2/6 as well as Battle-Fisher, M. (2011) Severity of scope versus altruism: working against organ donation's realization of goals—an essay. Online J Health Ethics 7(2). http://aquila.usm.edu/ojhe/vol7/iss2/4. Permission has been secured from the publisher.

A question to ask might be the moral entropy brought out when the values of members of a social network differ. More likely than not, everyone will never agree. In addition, ethical values are transitory and have gradients of buy-in. What can policy do to furtherance buy-in? "Potential" has its distinct silos in organ donation. As a *potential donor*, Battle-Fisher (2010) originally classified possible donors into three categories:

1. A "potential" with viable organs to agree and then act as donor
2. A "potential" with viable organs to choose not to act now (which the hope that this could change over time) but not against the idea, or
3. A "potential" with viable organs that is unobtainable with negative ethics toward donation

Upon revision, four updated variants of donor potentiality now account for the conservation and accumulation of feedbacks in social systems.

1. A "potential" that publically agrees with donation in all cases (pro-donation) and supports all donation options including becoming a possible donor himself.
2. A "potential" that publically agrees with donation in all cases (pro-donation) and but this kidney should come from someone else.
3. A "potential" that publically agrees with donation in select cases but not in the case of the patient in question (e.g. "I know mama won't take care of it so why bother", "Grandpa has lived a long productive life, why put him through this?")
4. A "potential" that has not committed to either pro or con publicly (perhaps in fear of reaction from others in the network)

An article in *Social Psychological and Personality Science* written by Nordgren and Morris McDonnell (2011) posed a research question that should be central to public health ethics. This was published in a psychology journal which may not be on the radar of many bioethicists. Nordgren and Morris McDonnell (2011) posit that rationality is thrown out of the window when the burden of people afflicted by a crime becomes incomprehensible.

The basic premises of "scope-severity paradox" according to Nordgren and Morris McDonnell (2011) are:

1. People only connect emotionally with crime victims within our personal social network (i.e. family, clan, neighborhood, civic group) that we know and care about.
2. Increasing the number of victims decreases the perception of severity of the problem. "Your problem, not mine" could be reworded as "call me only if a loved one is directly affected."

At a more elemental level, the lack of prowess in recognizing the gravity of an event would be explained by a presence of scope insensitivity. Desmentes et al (2007) note the importance of personal gain versus loss in unraveling this scope insensitivity. But public health must find a way of emphasizing a collective gain/loss framing that is linked to personal actions. What would be the social cost for donating (gain for society and personal gain for patient)? Moreover, is the public not paying in the end by the lack of living donations as both loss for society and personal losses for patients)? A central tenet of the Health Belief Model is perceived benefits where a health behavior must be framed as having a chance of affecting change in order to support the utility of that decision (Janz and Becker 1984). This would be an individualized framing of gain and loss. Does this cover all of bases of explaining the public's health? While perceived benefits are individualistic, what is lost is the exploration of the grand scope of the health concern on decision making.

Shared Loss for Public Welfare

Policy must reframe the public narrative to combat scope. According to the scope-severity paradox, the public does not easily care about those that people and narratives do not know, especially when they originate from masses of far-flung individuals. How much emotional energy can each realistically give, especially if specific energy needs to be directed to an unimaginable host of others? A lesson could be learned here in terms of framing scope-severity paradox around public health ethics. This is a prime example of the gain-framed experience that perhaps is radiating from a personal loss. Organ donation is a situation with the best intentions (altruistic compassion), but does not automatically resolve with a saved life in the end. Is the son gaining a mother by donation (gain-frame) or losing a mother by not donating (loss-frame)? Or is it both? As an "anticipated health behavior," living organ donation cannot be practiced and reinforced through active trial and error of the behavior (Battle-Fisher 2010). What can change would be the donor's belief as a potential donor, which is wedded to the possibility of ever-changing personal ethics (Battle-Fisher 2010).

The question, whether there can be sufficient saturation of altruistic compassion achieved in order to trigger innovation in a network, should be raised. After reviewing the legacy on donation prevalence after the policy was enacted, what happened to this health indicator under the policy in light of the system's history? Did the organ donation policy counteract the outflow of parameters draining any gains in organ donation? Or did the policy create inflow values that were protective of organ donation?

There is much to be said about defying the overall collective ethic of one's social network when the act for donation actually occurs and becomes embodied in convalesce and a physical scar reminding of the removal of "Uncle John's kidney" (Jones 2009). Unlike most illness discourses, there would not be physical changes to the body that the network would have to grapple warranting medical intervention. Most chronic illness is a deal breaker in living donation. So what is left is a negotiation of a possible medical event that can place an otherwise healthy individual in possible harm's way (though risks are miniscule). The potential donor is asked to play a role as a giver of organs and live to tell the story that end with a happy ending for another (Battle-Fisher 2010). Take Jones' (2009) well grounded point of the continued attachment of the transplanted organ to the donor. The network is consistently reminded of the divergent decision and may be called to help the donor when they disagreed with the initial decision. There goes the extemporaneous harmony.

The risk and rewards become that of the collective. Life no longer exists as an individual attribute but one that is negotiated with the needs and desires of the network in mind. The increased risk of chronic kidney disease and ESRD is outstretching the decreasing supply of viable kidneys available after each donation (assuming replacement that does not keep pace). But as time passes the physical body can only take so much wear and there will be a watershed "moment" when a loved one, such as Ann's cousin, needs an organ. The prevalence rates cruelly illuminate this possibility. People discuss certain issues with the Thursday night bowling league and a radically different set of bioethical discussions in Sunday school. But the public hold a quiver of stocked beliefs that are selectively shared. The nature of ethical deliberation is animated and changing. The most complex system is the one in which the public has the most to lose. Then it is up to human agency, clinical knowledge, and a network of gatekeepers as to an organ's fate. But as time passes, the state of the system moves on and policy must be created to account for the new fate as the public's ethical stances has evolved.

References

Battle-Fisher M (2010) Organ donation ethics: are donors autonomous within collective networks? Online J Health Ethics 6(2). http://aquila.usm.edu/ojhe/vol6/iss2/6. Accessed 14 April 2014

Berlinger N (2009) Friends in high places: doing bioethics at 36,000 feet. Bioethics Forum. http://www.thehastingscenter.org/Bioethicsforum/Post.aspx?id=3500&blogid=140. Accessed 14 April 2014

Berlinger N (2010) Spin doctors and torture doctors: inconvenient truths about complex systems. Bioethics Forum. http://www.thehastingscenter.org/Bioethicsforum/Post.aspx?id=4704&blogid=140#ix zz0sNaUFyK8. Accessed 14 April 2014

Desmentes R, Bechara A, Dube L (2007) Subjective valuation and asymmetrical motivation systems: implications of scope insensitivity for decision making. J Behav Decis Making 21:211–224

Janz N, Becker M (1984) The health belief model: a decade later. Health Educ Q 11(1):1–47

Jones N (2009) The importance of embodiment in transplant ethics. In: Ravitsky V, Fiester A, Caplan A (eds) The Penn center for bioethics guide to bioethics. Springer, New York

Nordgren L, Morris McDonnell M-H (2011) The scope-severity paradox-why doing more harm is judged to be less harmful. Social Psychol Personal Sci 2(1):97–102

Part II
Applications of Modeling to Health Policy

Chapter 8
Mental and Simulated Models in Health Policy Making

Essentially all models are wrong, but some are useful. (Box and Draper 1987)

Microworlds take our great ideas and mental hunches and input them into a "constructed reality" (Papert 1980). The formal system thinkers and their technologies share the policy glory as an understudy but they have to be in the race. In my opinion, system thinking does not take precedence over the art of policymaking. A system is inherent to the filigree of policy. It is a supporting role in assuring integrity in policymaking but system thinking should never be demoted to an understudy. Sometimes, we will not like the answer when the modeling, such as the offerings gained system dynamics modeling, upends our pacifying policy realities.

System Dynamics

System dynamics places "its emphasis on causal feedback as an organizing principle for explaining observed patterns of behavior" (Homer and Milstein 2004). System dynamics models are considered closed to external forces so that the rates internal to the system can be attributed to the workings of the social systems directly.

There are three distinct ways of visualizing system dynamics:

1. Simple causal loop diagrams (Fig. 8.1a)
2. Stock–flow (SF) diagrams (Fig. 8.1b)
3. Mathematical equations

Suppose that a policy is interested in getting to the structural knowledge behind elements of population change. Population change can be expressed in terms of shifts in mortality, fertility, and migration. First, the causal loop diagram has at its core two basic elements: behavior and structure. Looking at the analytical purpose of the causal loop diagram, such as the one illustrated in Fig. 8.1a, is to illustrate dynamic impact. There are the elements in the population model (population, births, deaths) are tied together by causal links. As it is well established that births and deaths change the population, all three of these elements are tied

© Springer International Publishing Switzerland 2015
M. Battle-Fisher, *Application of Systems Thinking to Health Policy & Public Health Ethics*, SpringerBriefs in Public Health, DOI 10.1007/978-3-319-12203-8_8

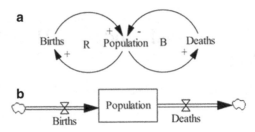

Fig. 8.1 Examples of a causal loop diagram of the cycle of population change (**a**) and basic stock and flow with one inflow and one outflow tied to population change (**b**). (Reprinted from BMC Health Services Research, 7, Elf, M., Putilova, M., von Koch, L., & Öhrn, K. Using system dynamics for collaborative design: a case study. 123. Using system dynamics for collaborative design: a case study. Open access. BioMed Central original publisher)

together by the arrows serving as causal links. Notice that the positive signs between population and births. The population procreates, therefore increasing the births (+) which subsequently adds (+) to the population. As for the effect of death, the population is the source of deaths, thus labeled as positive polarity (+). However, in an act of balancing the system, deaths take away from the stock thus labeled as negative polarity (−). It should be noted that additional elements can be added to the diagram. The causality, or better yet, how this system fits together with the additional restraints, may be far less apparent without using systems and modeling.

Stock and Flow of Public Health Interventions

The fluctuations in global populations have merited public health intervention. Another way to view this system is the use of the stock and flow diagram (as illustrated in b). The stock in the second figure is shown as the rectangle with valves (the coupled small rectangles on the pipes) that control the flow into and out of the stock, population. This gives a snapshot of a simple system that is the bedrock of understanding all social systems (see Gonzalez and Wong 2012).

> Unfortunately, there is increasing and robust evidence of a fundamental lack in the human understanding of accumulation and rates of change: a difficulty called the stock–flow (SF) failure. (Gonzalez and Wong 2012)

Stocks and flows are said to be the simplest dynamical system. Stocks must always be defined in terms of flows. A flow is akin to a rate of change over time (either inflow or outflow) of that stock. An inflow adds and an outflow depletes. Flows do not generate immediate changes in the stock to which they are connected. Thus is the requirement of time in order to understand the system. There can be any number of inflows and outflows and they do not have to be equal in number or rate of accumulation. Flows accumulate in a stock under the "principle of accumulation."" But it can still trip people up; this is the SF failure.

A stock is an entity that can be added to or depleted by flows over time. As Sterman (2000) explained, "stocks [including beliefs] accumulate past events." The stock does not change in an instant. Vaccination rates will not reach their goal instantly. The stock is at the mercy of the trickles and deluges that come and go in the flow mechanisms. Therefore, the stock in question can only be changed in terms of the magnitude of the changes in the flows, with the rates controlled by the valves. But people often find it difficult to predict the nature of the effect of the rates of the flows (see Hamid 2009; Booth et al. 2000). It is paramount to figure out what stocks, after the flows are figured out with modeling of course, are modifiable. If the stock is modifiable with the policy, the inference of the workings of the systems must be a central part of the policymaking cycle.

Sometimes, the bathtub overflows, as there is a faster rate of inflow that greatly overcomes the rate of outflow in a time period. As policymakers it is paramount to not only note the flows but also break down the variation in the valves that control these flows into a stock. At other times, the tub is bone dry as a faster rate of outflow overcomes the rate of inflow. The take-home message of this systemic tale is that reliance on knee-jerk reactions and conclusions cannot explain the time dimension and the rate of flows into a stock under policy consideration (see Meadows 1999).

Unmasking the goal of the system is the most "crucial determinant of a system's behavior" and the far too common practice of cherry-picking the lowest hanging fruit of a system in articulating the goals of the policy. Policymaking motivated by the lowest hanging element is the less effective strategy as the remainder of the "system" is still there and working (Meadows 2008). The singular element may not be the linchpin to the solution at all or if it is the element is cranked in the wrong operational direction. The perception of easy access to a particular target within a policy may not result in the positive results intended with that selection. Places to intervene in a system are not intuitive.

Meadows (1999) offered 12 leverage points that optimally affect a system. How Meadows brought these leverages to life is the story of a legend. These leverage points were shared extemporaneously during a meeting of powerful policymakers to advance policy. Wrong choice or ill use of appropriate choices impedes desired operational change. The excerpt below presents in her words admonishing policymakers to work with the nature of a system and not against it. She called for understanding the system which may not be the most politically expedient path. Meadows, in her unscripted walk up to a blank board during a global trade meeting, announced her hierarchy of leverages. She spoke to those with the power to affect the calculus of the trade policy to bar from making decisions based on elements that work against the health of a system.

> So one day, I was sitting in a meeting about how to make the world work better—actually it was a meeting about how the new global trade regime, NAFTA and GATT and the World Trade Organization, is likely to make the world work worse. The more I listened, the more I began to simmer inside. "This is a HUGE NEW SYSTEM people are invent-

ing!" I said to myself. "They haven't the SLIGHTEST IDEA how this complex structure will behave," myself said back to me. "It's almost certainly an example of cranking the system in the wrong direction—it's aimed at growth, growth at any price!! And the control measures these nice, liberal folks are talking about to combat it—small parameter adjustments, weak negative feedback loops—are PUNY!!! (Meadows 1997)

Meadows (1999) listed on the flip chart of this meeting the following leverages, from least effective (denoted as #12) to most effective (denoted as #1) places to target in affecting the nature of a system. These leverages are:

12. *Elements of a sociopolitical system (better known as the parts)*—It is the basis of debunking reductionism in affecting systems. Sometimes, the best systemic choice is not the call of the vocal public who by nature are not privy to the complexity inherent to making that choice.
11. *"Buffers" in relationship to its stocks*—Large stocks that often are too cumbersome or expensive to change.
10. *Actual structure of the stocks and flows or nodes of a network which may be unchangeable or too expensive to attempt to change*—After the structure is there, the power lies in acknowledging the missteps and mistakes in design.
9. *Attacking delays in the system "relative to the rate of system changes"*—"Things take as long as they take."
8. *Balancing Feedback Loops*—This negative, balancing loop, and its strength are there to keep focus toward a policy goal. They are necessary. We all use balancing feedbacks as a mode of controlling our environment. As a consequence, when a balancing loop leads a system astray, there must be "a goal, a detecting … device … and a response mechanism." This is the policy equivalent of putting out fires but there is always dry brush out there.
7. *Reinforcing Feedback Loops*—Unlike balancing loops, a positive, reinforcing loop gains momentum in growth and erosion of a system. Any "unchecked" reinforcing loops leads to the implosion of a system. When you slow a reinforcing loop, the system sprouts balancing loops.
6. *Leveraging Information Flows*—This requires finding out who is in the loop and who is out. Who in the network is in the know? Who in a social network has the influence to persuade and diffuse information?
5. *"The Rules of the Systems"*—This addition should make ethicists happy. This would be the rules of the collective and the individual—paternalism, autonomy, norms, values, ethics, laws, edicts, constitutions. Meadows added the distinction that in opposition to physical rules, social rules are "progressively weaker" with a slippery slope.
4. *Engaging changes to the self-organization of the system*—This is a largest predictor of resilience of a system. Giving a system the freedom to change requires a faith in experimentation that may be feared.
3. *Understanding the overarching goal of the system*—Analogous to a mission statement, "what is the point" of all of this?
2. *Paradigm Shifts*—While there is no monetary or physical cost to changing a mind-set, this may be the most bruising to a policymaker's ego. Business as usual may need to become history for the sake of effecting systemic change.

1. *Debunk the marriage to a paradigm in the first place*—Meadows called for nihilism in worldview. The goal would be to elevate beyond the Fox and certainly leave Hedgehog out of contention altogether. I see this call as most likely unobtainable level of self-actualization. But Meadows did frame this as a kind of striving to "Not Knowing." But the reality of policy is accountability to some sociopolitical frame. I am not sure if a policymaker can "listen to the universe" when health is at stake and some public is yelling in your ear for change. It would be a good job if you can get it and be able to keep the nihilism.

The Role of Simulation in Policy Making

Simulation's Gift to Policymaking

Justin Lyon, the founder and CEO of SIMUDYNE offered insight on the power of gamification as a health policy tool. As a mission of SIMUDYNE, simulations are at the core of supporting well-informed decisions of their clients. The mantra of SIMUDYNE in its use of policy simulation is to test scenarios before investing on time, energy, tangible resources, and cognitive power. Lyon entertained questions about the place of simulations in policy and the future of gamification as a mode of visualization for policy matters. According to Lyon, a problem in translation of mathematical models (which is the means to a policy end) is resolved in the approachability in using artificial life visualization. Lyon said that this kind of visualization "frees people to hold (policy) dialogue and debate without having to concentrate on backdoor mathematics." The nuts and bolts science of simulation is inaccessible to most, however, gamification in the form of avatars instantly makes the math accessible without being off putting. If the math is approachable, it is more likely to engage as a part of the policymaking process.

What power does simulation hold for health policy? According to Lyon, simulations serve to "reduce risk and reduce cost." But the financial, social, and spiritual costs attached to health care are mountainous. That vastness of systemic effect should by nature call for an investigation with new, dynamic eyes. Case in point, in terms of health promotion and intervention, Lyon offered that cautionary tale must be aware that health cannot shift out cost by concentrating on one specific change in one sector of health without realizing that financial and social cost shifting to another part of the system. What does that cascade of change look like? Graphs are static. Simulation can change and morph to suit new situational parameters. Choose simulation when rigorously conducted as well as pertinent to the policy matter being debate. The simulation, if used appropriately, could help to break the stalemates that can impede policy development (see Repenning and Sterman 1999; Sterman 2006; Sterman 2000).

> Regardless of the policy model used in a policy system, when a new policy is written or committed in the system, the administrator must consider [policy ratification as] how the new policy interacts with those already existing in the system. (Agrawal et al. 2005)

Practice Dynamics in Health Policy

Minyard and her colleagues at Georgia State University demonstrate that system thinking is not an esoteric, inaccessible concept but rather a set of skills that anyone—especially policymakers—can benefit from learning about and practically applying. In 2008, the Georgia Health Policy Center (GHPC) began an intensive educational program for state policymakers and legislative staff wanting a deeper understanding of health policy issues. The Legislative Health Policy Certificate Program (Certificate Program) is a continuing education program for state legislators and staff designed to build systems thinking competencies using health policy content. Two sessions address "core" health policy topics, such as health financing, insurance coverage, and access, while two sessions are devoted to issue-specific topics determined by the participants, ranging from childhood obesity to trauma care (Minyard et al. 2014). However, they believed that policymakers could also benefit from building their skills to approach policy issues as "system thinkers." That is, to look at the big picture, integrate diverse perspectives, consider changing dynamics, and explore high-leverage interventions in order to begin to change the way they frame issues, ask questions, and consider solutions to challenging health issues (Minyard et al. 2014).

After searching the literature for how research influences policy and how policymakers take in and respond to information, GHPC found that most education for legislators focuses on specific topics or supports specific policy decisions and that this approach was inadequate to generate the type of policy we—and policymakers—desired. System dynamics (and system thinking) provides a useful lens for approaching challenging policy issues. System thinking utilizes multiple disciplines and critical thinking skills such as dynamic thinking (looking at an issue over time), system-as-cause thinking, and forest thinking (looking at the big picture and how things fit together). This was just the type of learning capacity GHPC thought policymakers would benefit from and decided to build their curriculum around this approach (Minyard et al. 2014).

Simply asking questions, such as these, can dramatically raise the level of conversation.

- What does an issue look like if expressed as a trend over time? What might be causing that trend? Is there a systemic structure in place causing (or influencing) that trend?
- What policy might address this underlying trend?

- Are there potentially unintended consequences that could occur as a result of this policy?
- How long might it take to see measurable improvements in desired outcomes?

Richmond's conversational use of skills was used to create a six-question framework for policymakers to use when difficult policy challenges are encountered (Minyard et al. 2014).

Using this six-question framework, legislators were asked to draw behavior over time graphs to think about an issue not as *a point in time* but rather as dynamically *changing over time*. Throughout the course, these graphs were used to expand legislators' thinking on the health policy issue at hand. For example, when discussing rates of uninsurance in the state, looking at a static figure of the percentage of uninsured in year X gives only one picture of the problem. By expanding the time boundary and looking at uninsurance rates over a period of 30 years, one gets a more complete and nuanced picture of the problem (Minyard et al. 2014).

Other system tools used in the course were stock and flow maps, built in "real time" in the room to help illustrate the system and opportunities for high-leverage interventions. In one of the classes, GHPC built a prevention and treatment stock and flow map to highlight how the uncontrolled "flow" of healthy people becoming at-risk and then chronically ill creates a reinforcing loop of spending on one aspect of the system. Legislators had lively conversations over this map, discussing how not investing in prevention could get very costly down the road.

Another tool used in the course to help build system thinking skills was the development and use of simulation models. In 2008, participants in the first Certificate Program chose childhood obesity as an issue on which they wanted to learn more. As a part of the course, GHPC convened 15 Georgia legislators and staff for a half day and gave them a laptop with a proprietary simulation model based on a similar tool developed by the Centers for Disease Control and Prevention (Homer et al. 2007). The activity occurred in a real-time, hands-on learning lab environment where participants were encouraged to express assumptions, predict outcomes and inquire into differences between the two (Minyard et al. 2014). The model provided participants with an opportunity for a rich, lively hands-on conversation that would not have been as productive had it been conveyed only in a presentation form.

Reflective and Systematic Approaches

When people brainstorm, we battle with mental, qualitative models pieced together as divergent points (inputs). We have all been there. Mentally, we approximate all the time. It is a natural part of the policy process. To the miasmic stench of permanent markers littering a flip chart, the decision makers and perhaps a silent minority are left to make sense of that complex data to get at the process that has no obvious end in sight. There needs to be "reflective conversation" with the situation of the simulated results (Schon 1992). But the model alone will not be a predictive crystal ball. Much chatter has been circulated about the overreach of models into the world of prediction in policy.

The model does not forecast. The model may not be able to capture all of the connections to other policies that it is related to. It is a simulation. It is a model that is based on decision rules. It is a system with a priori boundaries. Systems offer math to support or refute initial reactions to early conditions viewed under a policy (see Forrester 2007). This is indispensable and this fact should not be taken lightly.

Each policy must be approached and respected as a necessity to prudent, systemic action to debate, not as a letdown of the expected policy desired beforehand. The satiation comes in what policymakers do with the systemic knowledge that is gained and ideally apply with prudent judgment and systemic consideration.

References

Agrawal D, Giles J, Lee KW, Lobo J (2005) Policy ratification. Policies for distributed systems and networks, 2005. Sixth IEEE international workshop on policies for distributed systems and networks, 223–232. doi:10.1109/POLICY.2005.25

Booth Sweeney L, Sterman JD (2000) Bathtub dynamics: initial results of a systems thinking inventory. Syst Dyn Rev 16:249–294

Box GEP, Draper NR (1987) Empirical model building and response surfaces. Wiley, New York

Forrester J (2007) System dynamics—the next fifty years. Syst Dyn Rev 23(2/3):359–370

Gonzalez C, Wong H (2012) Understanding stocks and flows through analogy. Syst Dyn Rev 28(1):3–27

Hamid TKA (2009) Thinking in circles about obesity. Springer, New York

Homer J, Milstein B (2004) Optimal Decision making in a dynamic model of community health. Proceedings of the 37th Hawaii international conference on system science. http://citeseerx. ist.psu.edu/viewdoc/download?doi:10.1.1.135.3301&rep=rep1&type=pdf. Accessed 15 April 2014

Homer J, Hirsch G, Milstein B (2007) Chronic illness in a complex health economy: the perils and promises of downstream and upstream reforms. Syst Dyn Rev 23(2/3):313–343

Meadows D (1997) Places to intervene in a system. Whole earth review. http://www.wholeearth. com/issue/2091/article/27/places.to.intervene.in.a.system. Accessed 8 June 2014

Meadows D (1999) Leverage points: places to intervene in a system. The Sustainability Institute, Hartland

Meadows D (2008) Thinking in systems: a primer. Chelsea Green, River Junction

Minyard KJ, Ferencik R, Phillips MA, Soderquist S (2014) Using systems thinking in state health policymaking. Health Systems. doi:10.1057/hs.2013.17

Papert S (1980) Mindstorms. Basic Books, New York

Repenning N, Sterman J (1999) Getting quality the old fashioned way: self-confirming attributions in the dynamics of process improvement. In: Scott R, Cole R (eds) The quality movement and organizational theory. Sage, Newbury Park

Schon D (1992) The theory of inquiry: Dewey's legacy to education. Curric Inquiry 22(2):119–139

Sterman J (2000) Business dynamics—systems thinking and modeling for a complex world. Mc-Graw-Hill, New York

Sterman J (2006) Learning from evidence in a complex world. Am J Public Health 96(3):505–514

Part III
A Brief Exploration of the Complexity of Health Disparities (As Humanistically as Possible)

Chapter 9
Social Disparity, Policy, and Sharing in Public Health

Race, Gender, and Social Interaction

Social interaction is a byproduct of being a part of a (social) system (Giddens 1979). Let us take for instance, the relocation of disadvantaged populations into areas thought to bring more chances for opportunity. The work of Marsden (1987) supports the notion of highly homogeneous support within American society. Individuals often chose to associate with people like them and report that others in our networks do the same (Marsden 1987; McPherson et al. 2001; Louch 2000). But choice may be forced. Housing laws do not often result in a reversal of homogeneity. A case in point is the policy issue of housing desegregation and the social ties that bind black families.

A disparity essentially means that in the whole scheme of things, some individuals that tend to share certain characteristics are left behind. People are both bolstered and constrained by the very society in which they exist. It is wise investing our policy efforts that leverage the most return for limited resources and effort (Resincow and Page 2008; Meadows 1999). This is if that leverage point is capable at all of satisfying the social return at all. Policymakers must choose wisely.

> It is not the intelligent woman vs. the ignorant woman; nor the white woman vs. the black, the brown and the red,-it is not even a cause of woman vs. man. Nay, 'tis woman's strongest vindication for speaking that the world needs to hear her voice. (Cooper 1988)

> Only the BLACK WOMAN can say, when and where I enter, in the quiet, undisputed dignity of my womanhood, without violence and without suing or special patronage, then and there the whole Negro race enters with me. (Cooper 1988)

> Our reactions are far from linear. We are human. Black women may converse in the salon, free to vent among the miasmic heat of irons, stench of acetone nails and lye-laced chemically treated hair. The conversation mixes purpose with the unbridled freedom to discuss her life, love and sorrows. This self may not be demonstrated in the clinician's office but the same sorrow song of her life is necessary to care for her health. Looking at the health disparities plaguing the U.S., the self-projection by the Black female patient must be unconstrained and she must "speak" to respectful clinicians in return. (Battle-Fisher 2013a)

© Springer International Publishing Switzerland 2015   77
M. Battle-Fisher, *Application of Systems Thinking to Health Policy & Public Health Ethics*, SpringerBriefs in Public Health, DOI 10.1007/978-3-319-12203-8_9

For a black woman, she is her social effeminacy. Playing race and gender requires making a conscious determination of the internalized and socially projected selves. Research has shown stark disparities in health outcomes for black women in the USA across all ages and socioeconomic backgrounds. Phelan, Link, and Tehranifar (2010) cite the persistent health disparities have proved resilient to most health promotion efforts. Recalling Link and Phelan's theory of fundamental causes, this chapter is a call to dig beyond the surface to uncover mechanisms perpetuating disaparties (see Phelan et al. 2010). Determining "when and where I enter" in light of resource disparity may be exasperated by intervening mechanisms set to improve outcomes in health interventions (see Phelan et al. 2010). An issue of embodiment is a precursor to policy development that must not be ignored. Epidemiologically, each woman has a footprint of exposure over the lifespan that bolsters or hinders realizations of health. Sex matters in etiology of disease. Gender is linked to life chances.

The Choice Between Two Overlapping Doors

Dissatisfied with the gender-based prejudice, Cooper spoke of making a choice between two rooms at a hotel. One was labeled "for ladies" and the other "for colored people". She mused "under which head it come" (Cooper 1988). Society often places race above all other social constructions. Society had dictated to which she must oblige "colored". Living a departmentalized self for the sake of social stratification serves no one. But the reality remains that it is this stratification from which we tend to self-define or get "self" defined through proxy. Although most would now agree that race is socially created, the dark hue of one's skin does not come with a disclaimer that the hue should not matter. Cooper highlights that society often demarcates based on social dichotomies (Cooper 1988). In doing so, a woman must contend with whether she is black or not, or healthy or ill.

> The history of the American Negro is the history of this strife- this longing to obtain self-conscious (person) hood, to merge (the) double self into a better and truer self. (DuBois 1965)

Multiple, layered selves are engineered yet are socially and physically coped with at varying degrees of success.

> With identity thus re-conceptualized [with intersectionality], it may be easier to understand the need for, and to summon the courage to challenge, groups that are after all, in one sense, 'home' to us, in the name of the parts of us that are not made at home. (Crenshaw 1994)

Would a black woman then be embodied as "female" not black if she has successful beat the grim statistics of black morbidity and her progeny? When can she be black again? Is she then something else? Anna Julia Cooper wrote in 1892 that there are two kinds of peace [balance]in the world: one produced by suppression and the other brought by "proper adjustment to living, acting forces." (Cooper 1988). "Other-isms" are determined inside (endogenous) or outside (exogenous) the social system.

Just as a system variable leaves behind a history, so does the veiled history of people of color (see Sterman 2000; DuBois 1965). If the peace allowed by optimal health remains confined to the uncertainty of such a hermeneutic, not exploring "the Other" of the black female or any other marginalized groups, for that matter, may be the worse folly of all, in policy. Even if race, gender, or take your pick is found to be exogenous, it does not mean that this striving to reconcile dichotomies is inconsequential. As written by Harvey (2001), complexity itself can be "socially determined, productions of historically situated social structures". Social complexity in networks based on "other-isms" requires the critical examination of the overlapping multiplexity of roles that are affected by social constraints (Crenshaw 1994; Verbrugge 1979). Roles, while important in understanding the nature of social relationships, are not the same as social projections of worth both internally and externally ascribed to those roles.

> Community is ingrained for some. For others, they live only in a zip code. (Battle-Fisher 2013b)

> Our modern world may shrink due to common interest while the physical and emotional resources necessary to house us become more and more constrained. Such constraints can place undue burden on the state of health among urban dwellers. (Battle-Fisher 2013b)

Housing and Equal Opportunity Impacts

Schelling's (2006) groundbreaking work in racial segregation is often called *Schelling's Tipping Model* in his book *Micromotives and Macrobehaviors*. Briefly, Schelling's segregation model intended to discern the effect of spatial proximity on racial and economic segregation. There are instances when rational agents help to clarify without the whirlwind of cognition, such as in instances of game theory. This model is a rule-based agent-based approach to understanding social segregation. It may be further discerned that if there is segregation that this may be framed as a form of status homophily that has been encouraged by *de facto* and *de jure* mores and laws of racial regulation. Birds of a feather endure disparity together. As an agent-based model that established an underlying mathematical rule to represent agent behavior, the Tipping Model used the neighborhood as the unit of analysis. The decision rule was whether the rich or poor decided to stay or move. Three components of his model are:

1. The agent—person in the neighborhood with the ability to decide whether to stay or move
2. The rule as the percentage of threshold tolerance (e.g., 30 % similarity wanted by an agent)
3. The aggregate behavior of all agents based on the percentage of similarity desired and required to make informed decision of leaving.

The selection of these components was important not only to assure the integrity of the model but also demonstrate the macro/micro levels in decision making. Macro level meant segregation. Micro level in the model translated to the level of accepted tolerance for social diversity.

Schelling (2006) found a *threshold* point that people can tolerate until that point becomes undesired. In order for Schelling's model to be completed, balanced mathematically, the model exhibits too much randomness to even be achieved. Schelling created "happiness (mathematical) rules" for expressing threshold-based preferences toward segregation (Schelling 2006). If an agent's tolerance for diversity was low in accordance to the mathematical rules. the agent would flee in the model. Racial tolerance was defined mathematically as being 'happy' but staying put (Schelling 2006). But his model was still based on rules. Practically, these outcomes translate to everyone in that neighborhood being happy with their personal welfare with the residential mix based in part on the mobilization efforts. Living as one is sometimes saccharine.

When an individual under peer influence duplicates the exact action of another (such as leaving a neighborhood), this is called an exodus tip that occurs due to a departure from a system. If one person moves into the system and another leaves that system, this is known as a genesis tip. No neighborhood can sustain a balance of equal representations based on a social attribute such as race. The index of dissimilarity shows the extent of that imbalance. Card, Mas, and Rothstein (2008) found, using US Census Track data from 1970 to 2000, that the threshold or tipping point for white flight was higher for communities that exhibited a higher race-based tolerance. The tipping points ranged from "5–20% minority share" (Card et al. 2008). Where does this white flight tipping point leave urban revitalization projects?

Gentrification policies move individuals with financial means into often income-depressed neighborhoods in order to lift up and bring back the neighborhood. An approach that does not account for the complete actions of agents regrettably underestimates the dialectic of connection, choice, and social function. Does the Schelling model still apply in the same way today? How is this tipping affecting the framing of urban housing policy? Policies can help to move bodies into shared social spaces. But those bodies with means are ambulatory. Lack of opportunity clips wings hampering a flight of another kind. Aspiration is often trampled by poverty.

A child that lived in the now-demolished Cabrini Green formerly splicing Chicago's Near North Side sky view spent years co-existing, and surviving with people that loved or loathed each other. How does that child with a Cabrini identity then leverage finding and maintaining beneficial ties outside of that Cabrini space in a gentrified neighborhood? Distance apart may not make the heart grow fonder. As previously discussed, homophily is a social characteristic describing the tendency to share one or more attributes. In terms of social and racial disparities, someone may gravitate to someone for knowing his pain.

Marginalized groups are often inundated with social hazards buoyed by diminished chances of physical migration (leaving). Therefore the structure of the social network may become more salient. A person is less likely to flee undesirable social circumstances without the social and financial resources to do it. Also a person for the sake of keeping support nearby may choose to forego leaving. Individuals with the fewest ties to reach a person or who shared a social attribute were found to be effective in supporting homophily (Kossinets and Watts 2009).

Policy Affecting and Targeting Underserved and Vulnerable Populations

Why is it so important to account for resilience of social networks when drafting policy that targets underserved populations? Cornwell and Waite (2009) found that social disconnectedness and perception of isolation were associated with lower reports of self-rated physical health. Further exploring of the influence of social ties on health, Cacioppo and Hawkley (2003) stressed that heightened susceptibility of isolation is of consequence among the aged and minorities. If policy was able to foster supporting ties over the lifespan, there could be an opportunity to counteract stressors and improve Quality of Life (QOL) of those targeted (and consequently those around them). Fischer (1982) found that the configuration of the kin (family as self-defined) and non-kin within a social network is pivotal to understanding the dynamic of information diffusion and influence. Bott (1957) noted that the person's "community" is not location based but rather is the network of social ties maintained by the families. I would argue that still stands today. We look out of our windows and see the effects of propinquity.

Ajrouch and colleagues (2001) specifically found that minorities had less diffuse, denser networks so more possible ties are utilized. In addition, blacks had more family members in their support systems and younger subjects had a larger proportion of kin than comparable whites in the study (Ajrouch et al. 2001). This finding highlighted that the concentrated dense "first zone" connections are pivotal as each connection may hold more influence over the decision making of the person. Negative and positive health effects have been found in social support. Cornwell and Waite (2009) found that social disconnectedness and isolation were associated with lower self-rated health. Lin (2000) reported the connection between social support and ties noting that social resources are "embedded in the ties of one's [social] networks". Support affects everyone, albeit in different ways. The quality of ties is how you get over.

How could policy even tackle such a hulking task? If the Schelling model still applies in the same way today, how is this tipping affecting the framing of urban housing policy? Policies can help to move bodies into shared social spaces and blend social realities. The citizens do not have to bond though sharing a block of beautified brownstones. Housing policy often targets co-location of bodies but the capacity to bond as a community can torpedo even the mosy well-intended policy. Those who are transplanted may have new support to thrive and become upwardly mobile. Or the din of disconnection can be deadening.

It Is Not the Side Effects' Fault

Sterman (2000) warned us to not become complacent by the faulty belief systems that worship the quest for "side effects" and their presumed effects on policy. According to Sterman (2000), a realized effect is "just an effect", not a "side effect". The dynamic "effects" are either:

1. Main or intended effects—the good ones that the policy intended in the first place
2. Signs of an "understanding of a system that is narrow or flawed" leading to undesired ones (mistakenly called side effects). (Sterman 2000)

Some policies could possibly be adequate if enacted true to spirit resulting in desired main effects. Unsuccessful effects and feedback undermine the original intention of the policy. It is agreed that policies must be well crafted and actionable but sometimes effects crop up that come out of left field. The landfall of the effects of a policy does not always match up with the anticipated outcomes upon which the policy was based. Sterman (2000) contended that with complexity comes a long-time horizon from cause to effect.

> Failure to recognize the feedbacks in which we are embedded, the way in which we shape the situation in which we find ourselves, leads to policy resistance as we persistently react to the symptoms of difficulty, intervening at low leverage points and triggering delayed and distant, but powerful feedbacks. The problem intensifies, and we react by pulling those same policy levers with renewed vigor, at the least wasting our talents and energy, and all too often, triggering an unrecognized vicious cycle that carries us farther and farther from our goals. (Sterman 2002)

Case for Your Consideration: Urban Housing and Development in the USA

The US Department of Housing and Urban Development (HUD) reframed urban revitalization as a 10-year Section 8 voucher Moving to Opportunity (MTO) program in New York, Los Angeles, Chicago, Boston, and Baltimore that transplanted lower SES families into more affluent neighborhoods (see Feins et al.1996). Interested in the main effects that such migration may have on the adolescents of volunteer families relocated under the vouchers, Kessler et al. (2014) conducted a randomized study with three groups:

1. Experimental group 1 with voucher, counseling, and adjustment assistance
2. Experimental group 2 with voucher and no additional support services
3. Control group with no voucher

Network research has found strong evidence that lower-rated health has been tied to homogeneity and bond formation. Young people tend to seek the support of peers.

The intended main effect of the policy was to expose lower-income families to increased social capital and opportunity. Young men were found to be adversely affected by depression, posttraumatic stress syndrome, and conduct disorder in both experimental groups when compared to the control (Kessler et al. 2014). But here is a lesson for the policy. Girls were found to be far more resilient and the moves were statistically protective. Development of a policy may have differing effects on the collective family unit as well as the welfare of the individuals themselves. The authors stated that "it is difficult to draw policy implications...policy will have to grapple with this complexity based on the realization that no policy decision will have benign effects on both boys and girls" (Kessler et al. 2014).

Questions for Discussion

1. How do you quantify social effects for public policy?
2. What research do you choose to base policy?
3. Should we go with the propensity of evidence under policy of old or new evidence that may better reflect policy under the changed social landscape?

Tama Leventhal and Jeanne Brooks-Gunn (2003) performed the first rigorous randomized control findings exploring neighborhood effects to mental health in a randomized control trial of New York City volunteer Moving to Opportunity (MTO) families. Unlike Kessler et al. (2014), Leventhal and Brooks-Gunn (2003)'s research with a 3-year follow-up found boys between 6–12 to self-report fewer depressive issues which may "result from (the boys') ability to travel back to their old high-poverty neighborhoods or from disruption of peer networks, which are salient during adolescence". In reexamining MTO data, even a slight improvement in "opportunity" into low poverty areas was protective for both boys and girls (Leventhal and Dupéré 2011). But a trigger switches in adolescence. Mental health worsens in the teen years when coupled with neighborhood mobility. In a "realist" evaluation of MTO programs, Jackson et al. (2009) warned of resisting the allure of "what works" without fighting to unearth the context in which the housing policy lies.

> Without people and human relationships, there is no neighborhood—there is simply a physical place. (Jackson et al. 2009)

Assessment and Evaluation of Health Policy

"Things take time" in policy to assess success (see Meadows 1999). What systems policy offers is not the advisement of the current policy decision, "but rather on how to change policies that will guide future decisions" (Forrester 2007). Ethically and politically, how much time is comfortable for policymakers to let things play out? The power in using systems to examine policy lies in its ability to offer insight on how a policy decision made now might cause systemic ripples based on the changing social conditions (Forrester 2007).

The structure is already present as the voucher program was already in place and had already started its effects on the families as well as the affected neighborhoods. Intact families move lives together but members' lives in effect can take different trajectories. The house is a home perhaps only for some. The fabric of the neighborhoods tipped. Systems thinking teaches that every system at the local level may not mirror the expectations of effects at the global level. An element at work in a system cannot and must not be divorced from the larger effect that element presents on the integrity of the system. What we have to consult is the current state of the research. As the social tides change, we have changes flowing in and among overlapping elements and systems.

First, policymakers must honestly assess whether the policy indeed leads to the effect that is wanted in the first place for the populations targeted as a population as well as local collectives (see Sterman 2000). Models from Kossinets and Watts (2009) additionally found that "forward looking" individuals have a greater penchant to get closer to people that they want in their network. But what forward thinking could marginalized populations call upon when social capital is bankrupt? I contend that the some strictures on civil society constrict people unapologetically. Bourdieu places the ability to have power gained over one's personal situation (social capital) squarely into networked relationships (Bourdieu and Wacquant 1992). The "habitus", as coined by Bourdieu, is a person's subjectivity of experience which takes place in the "field".

Networks support social capital or share the lament. Portes and Sensenbrenner (1993) argued that social capital must be redefined, as "expectations for action within a collectivity". The parameter often targeted by urban policy is to move people to a new land of opportunity—a new abode, a new school, a new job with promises of a better life. It is painted as removing the underprivileged to a public space askew in its allowance of economic and social opportunity. What of the bodies that are being moved? While systems have antiquity that lingers, policies have a legacy that leaves its own residue of feedbacks behind.

References

Ajrouch K, Antonucci T, Janevic M (2001) Social networks among blacks and whites: the interaction between race and age. J Gerontol B Psychol Sci Soc Sci 56(2):S112–S118

Battle-Fisher M (2013a) The heavy load of agency on health: Anna Julia Cooper, race and gender. Orgcomplexity blog. http://wp.me/p32x8n-q. Accessed 19 June 2014

Battle-Fisher M (2013b) Urban greenspace and collective health ownership. Mindful Nat 6(3):33–35. http://www.humansandnature.org/urban-greenspace-and-collective-health-ownership-article-159.php?issue=21. Accessed 14 April 2014

Bott E (1957) Family and social networks. Tavistock, London

Bourdieu P, Wacquant LJD (1992) An invitation to reflexive sociology. University of Chicago, Chicago

Cacioppo J, Hawkley L (2003) Social isolation and health, with an emphasis on underlying mechanisms. Perspect Biol Med 46(3 Suppl):S39–S52

Card D, Mas A and Rothstein J (2008) Tipping and the dynamics of segregation. J Econ 123(1):177–218

Cooper A-J (1988) A voice from the south. Oxford University Press, New York

Cornwell E, Waite J (2009) Social disconnectiveness, perceived isolation and health among older adults. J Health Soc Behav 50(1):31–48

Crenshaw KW (1994) Mapping the margins: Intersectionality, Identity Politics, and Violence Against Women of Color. The public nature of private violence. (eds) Martha Albertson Fineman and Rixanne Mykitiuk. Routledge, New York:93–118

DuBois, WEB (1965) The souls of black folk. In: Franklin JH (ed) Three Negro classics. Avon, New York

Feins JD, Holin MJ, Phipps A (1996) Moving to opportunity for fair housing program operations manual. Abt Associates, Cambridge

Fischer C (1982). To dwell among friends: personal networks in town and city. University of Chicago Press, Chicago

Forrester J (2007) System dynamics—the next fifty years. Syst Dyn Rev 23(2/3):359–370

Giddens A (1979) Central problems in social theory: action, structure, and contradiction in social analysis. University of California Press, Los Angeles

Harvey D (2001) Chaos and complexity: their bearing on social policy research social issues. http://www.whb.co.uk/socialissues/harvey.htm. Accessed 15 April 2014

Jackson L, Langille L, Lyons R, Hughes J, Martin D, Winstaley V (2009) Does moving from a high-poverty to lower-poverty neighborhood improve mental health? A realist review of 'moving to opportunity'. Health Place 15(4):961–970

Kessler R, Duncan G, Gennetian L, Kling J, Sampson N, Sanbonmatsu L, Zaslavsky AM, Ludwig J (2014). Associations of housing mobility interventions for children in high-poverty neighborhoods with subsequent mental disorders during adolescence. JAMA 311(9):937–47

Kossinets G, Watts D (2009) Origins of homophily in an evolving social network. Am J Sociol 115(2):405–450

Leventhal T, Brooks-Gunn J (2003). Moving to opportunity: an experimental study of neighborhood effects on mental health. Am J Publ Health 93(9):1576–1582

Leventhal T, Dupéré V (2011) Moving to opportunity: does long-term exposure to "low-poverty" neighborhoods make a difference for adolescents? Soc Sci Med 73(5):737–743

Lin N (2000) Inequality of social capital. Contemp Sociol 29:785–795

Louch H (2000) Personal network integration: transitivity and homophily in strong-tie relations. Soc Netw 22:45–64

Marsden P (1987) Core discussion networks of Americans. Am Socio Rev 52(1):122–131

Meadows D (1999) Leverage points: places to intervene in a system. The Sustainability Institute, Hartland

Phelan J, Link B, Tehranifar P (2010) Social conditions as fundamental causes of health inequalities- theory, evidence and policy implications. J Health Soc Behavior 51:s28–40

Portes A, Sensenbrenner J (1993) Embeddedness and Immigration: notes on the social determinants of economic action. Am J Sociol 98(6):1320–1350

Resincow K, Page S (2008) Embracing chaos and complexity: a quantum change for public health. Am J Publ Health 98:1382–1389

Schelling TC (2006.) Micromotives and macrobehavior. Norton, New York

Sterman J (2000) Business dynamics—systems thinking and modeling for a complex world. McGraw Hill, Boston

Sterman J (2002) All models are wrong: reflections on becoming a systems scientist. Syst Dyn Rev 18:501–531

Verbrugge L (1979) Marital status and health. J Marriage Fam 41:267–85

Chapter 10
The Concentric Model of Health-Bound Networks

Introduction

The roles become muddled with roles that support health and others that result from nonmedical reasons. It may be a foredawn conclusion that the person's quality of life (QOL) stays just that—inherent to a personal experience with disease. More-over, QOL is often analogous to length of life and prognosis. What if the patient believes that life is fine as it has been dealt, imperfect clinically but personally acceptable (or tolerated)? What happens when the self-care decisions of a patient run counter to the evidence-based prescriptivism of medical care? There is a blurring of the penumbras of the public and private spheres in our understanding of QOL. The complexity of social relationships and support requires policy for social integration has been shown to be linked to both physical and mental health. As the patient is embedded into a support network, I additionally posit that there is a "shared" collective QOL on the microlevel by which caring others are affected by the life state of the patient. It is an interesting question to explore the final victor in QOL: the psychometrically measured QOL which is constructed by the "others" in the medical establishment or a patient's subjective understandings of a sick existence. In light of generational issues of longevity with decreased physical and mental functionalities of patients, what must not be ignored is the network of support.

> A verbal model is better than no model at all, or a model which, because it can be formulated mathematically, is forcibly imposed upon and falsifies reality [...] It may be preferable first to have some nonmathematical model with its shortcomings but expressing some previously unnoticed aspects, hoping for future development of a suitable algorithm, than to start with premature mathematical models following known algorithms and, therefore, possibly restricting the field of vision (von Bertalanffy 1968).

Earlier versions of this chapter were presented at 2011 Aging and Society: An Interdisciplinary Conference, University of California, Berkeley, CA, and the 2012 Aging and Society: An Interdisciplinary Conference, University of British Columbia, Vancouver, Canada.

© Springer International Publishing Switzerland 2015
M. Battle-Fisher, *Application of Systems Thinking to Health Policy & Public Health Ethics,* SpringerBriefs in Public Health, DOI 10.1007/978-3-319-12203-8_10

Social relationships, or the relative lack thereof, constitute a major risk factor for health—rivaling the effect of well-established health-risk factors such as cigarette smoking, blood pressure, blood lipids, obesity, and physical activity. (House et al. 1988)

Quality of Life (QOL), Disability, and Autonomy

Pervasive poor QOL marks society's as a measure of the state of the public's overall health. There are social and economic implications to acting as a paternalistic ward of the nation's health. The concepts of *personal autonomy and collective autonomy* at first blush appear to be incongruent at best. Healthy patients demand a "better" QOL as an inalienable right of getting treated. Do the concerned others also have some "say" in what QOL should and must be for the patient? The fact that the society has become so vocalized at midterm elections bodes the concern of where the line between personal autonomy and collective self-governance come together. Is there a duty (as an appendage of "autonomy") that asks for duty not to merely act in one's own interest but also that of the collective?

While autonomy is often presented as a hallmark of medical ethics, autonomy is best framed with degrees of variation. It cannot be ignored that there can be social influence from others on patients. Socially constructed autonomy differs from the vanilla version as there is an assumed outside influence on the QOL of a patient. The effect of collective autonomy is shown as the rights are afforded within a self-governing network. Take the case of a family that has been stricken with end-stage renal disease (ESRD). There may be apathy to the perceived benefit of proper self-care and the quest for the elusive QOL that the clinicians expound as the Holy Grail. Other family members have languished on dialysis and complied at varying degrees toward "better QOL" and still died. A young person ran away but returned. This emotive back story could have grave consequences on compliance within the person's network. So what is the use of complying with physician's directives or social services when the network expresses discontent with the patient's health choices?

Joanne Lynn (2014) of the Center for Elder Care and Advances Illness of the Altarum Institute wrote that in observance of US caregiver policy there is a misplaced "overinvestment in health care and an underinvestment in support services" in a *Journal of the American Medical Association* editorial. While policies have placed insufficient priority in abating the stress on caregivers, patients continue to age and caregivers continue to be overburdened. The caregivers often are regulated to piece together support mechanisms, often to the detriment of the caregivers' own social integration (Lynn 2014).

Disability compels engaged nodes in a support network. The demographic shift could result in an older, more infirm population that will have a more profound demand of caregivers. According to projections of The US Department of Health and Human Services, by 2050, the number of individuals over the age of 65 will double from projections in 2010 (Department of Health and Human Services n.d.). The policy must act to temper the burden of costs for medical care in light of the skyrocketing number of aging boomers around the bend. While those on dialysis aged 65 and over have fallen in absolute numbers, QOL and life expectancy falls severely as an ESRD patient initiating dialysis at 65 or older (Tamura et al. 2012; United States Renal Data System 2013)

Concentric Model for Health-Bound Networks

An exemplary model for the integration of federal oversight and nonprofit organization was created when US Congress passed the *National Organ Transplant Act* (NOTA) in 1984. The act established the *Organ Procurement and Transplantation Network* (OPTN) to maintain a national registry of organ allocation and lead the on-the-ground charge in attacking health disparities in ESRD. Unfortunately, not enough kidneys are available so patients turn to long-term caregiving. The reality is that the kidneys are not keeping pace. The patients get older. Policy must also account for the life on dialysis. People have to keep continue on with damaged nephrons. "Donate life" but ESRD patients press on, abiding by the rules and constrictions of health disparities dampening QOL. Reality becomes "donate" caregiving support so the ESRD patient can "live" waiting for a kidney. From a clinical standpoint, a kidney or even dialysis may not improve life expectancy and QOL, which are among the main reasons for the clinical interventions.

> We need primary care providers to facilitate health care decisions for the dyad—to think about each decision as it affects a 78-year-old man with Alzheimer's and how it affects his 74-year-old female caregiver. We need to think beyond patient-centered care to dyadic-centered care or even caregiver-centered care.
> (Rosenthal 2014)

The concentric model of health-bound networks (CMHN©) is a model that seeks to examine changes in social support for chronically ill patients based on established overlapping social network research. Society is not defined by isolate but is understood rather as a "network of (social) networks" (Everett and Borgatti 1998; Gao et al. 2012). CMHN© in Fig. 10.1 illustrates this reality as nested communities, labeled as "spheres" in this model (see Everett and Borgatti 1998; Palla et al. 2005). The overlapping networks proposed by the model are communities built in differing degrees to the patient's (see Newman 2003; Palla et al. 2005). The CMHN© consists of four "spheres" serving as overlapping communities with their own unique networks (networks within networks). Each sphere houses its own social network defined by circumstance and duty to the patient. They are: the health network (labeled specifically as a kidney network as illustrated in Fig. 10.1), the general well-being network, the social network, and the polis. A graphic of the overlapping communities or spheres can be found in Fig. 10.1 below.

Nesting consists of cross-level influences and relationships between and among levels including established norms, standards, and social networks, all attributes that authors postulate are central to a new conceptualization of QOL (Gregson et al. 2001).

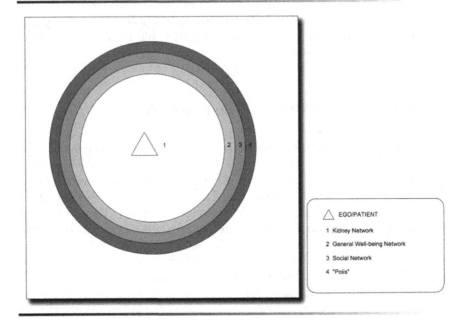

Fig. 10.1 The Concentric Model of Health-bound Networks (CMHN). CMHN visualizes changes in social support network composition. The ego (denoted by a triangle in the kidney network) is the patient in question. There are overlapping networks in the model's four spheres: the 'kidney' network (which may more generally be labelled the health network), the general well-being network, the social network, and the polis

The Health Network The health network is the core and most inner sphere of the model (Fig. 10.1). It should be noted that the exemplar in Fig. 10.1 is labeled "kidney network" allowing nominalization based on the illness narrative under investigation. This label would be open to change based on the illness being explored; for instance, the inner sphere may read "breast cancer network" as research has shown that QOL has been best measured as disease specific. The network revolves around the "ego" patient who is connected to other nodes. Specifically, members of this "kidney" sphere would include the ESRD patient (the ego), and socially invested alters (defined as caregivers or concerned, active others). Everyone in a network is interdependent in the mathematical and social sense. Remember that networks by definition embrace and account for interdependence both theoretically and mathematically.

As a strong tie or linked relationship, the nature of engagement could be more engaged and fortified within a network so there can be encouragement and trust among the members of the network (White and Houseman 2002). Carpenter, Esterling and Lazer (2003) found in a simulation model of health political networks that increased burden of involvement in the affairs of the networks depended most on strong ties. Strong ties can be socially depleting, requiring with five times the maintenance effort over weak ties (Carpenter et al. 2003). The engagement in cultivating such strong ties, which are often kinship based, is time and resource intensive (Granovetter 1973). What if there are instances where a strong tie could be more optimal for social support? Within this sphere where there is an expectation of high support, a person with a more casual interest (weaker tie) might prove harmful to the health of the kidney network. Is it just best to leave them be and let them sashay away or lumber away with a gait rattled by conscience? The answer in terms of chronic disease outcome management may not so simple.

The General Well-Being Network and the Social Network As illustrated in Fig. 10.1, the second most inner sphere, general well-being network, refers to a larger network including health network and active caregivers in life situations who may not exclusively be health related. The third most inner sphere is the "social network", a yet larger network which may also include probable and inactive caregivers. This would be the location of individuals with a more casual social role. But what will keep him or her there over the long haul?

The "Polis" Network Within the CMHN©, I refer to the polis in Fig. 10.1 as the "public" sphere or the general population affected by population health. The term "polis" is borrowed from the city–state of Aristotelian politic as it illustrated the expansive and intertwined nature of community formation (see Aristotle 1959). The polis is infinite in size but for illustration, the sphere is given an outside boundary in the model (see Fig. 10.1). Newman (2003) noted the futility of trying to capture representation of all possible members. What we will catch is what is named and available for inclusion in the analysis. The individuals exclusively found only in the outer ring of the polis may come to understand QOL as a politically driven or policy-based concern as opposed to an intimate discourse connected to the patient in question.

These spheres and their networks are possibly more meaningful over time. Because there will be losses due in the future, the effect of such loss can be captured mathematically. Real-world networks may be more resistant to the random loss of nodes from a network (Newman 2003). However, the tolerance of loss is allowable only to a point. For instance, Albert (2000), using World Wide Web simulations, supported the inherent resistance to random loss of alters. However, when actors were randomly lost at a rate as low as 7% within a single network, it proved detrimental to the health of the network (Albert et al. 2000).

How can the movement across networks be conceptualized? Two distinct processes could mark this movement across spheres and embedded networks:

1. Direct migration
2. Intermediary migration

Direct and intermediary migration within CMHN©, as illustrated in Fig. 10.2, visualized the movement as gain/loss of nodes across the spheres that occur over time:

1. Either through a direct ultimate path within a finality in destination (direct) or
2. Via a series of intermediate steps across spheres to a new support role (indirect)

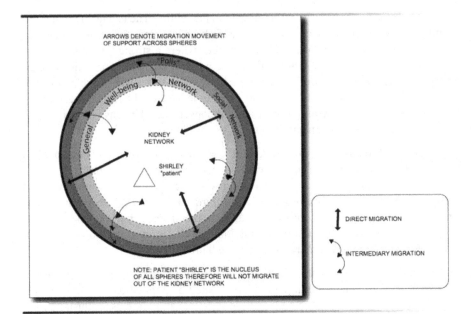

Fig. 10.2 Direct and intermediary migration patterns in CMHN. For illustration, the ego is 'Shirley' denoted by the triangle in the kidney network. The arrows (explained in the legend to the right) visualize possible patterns of gain and loss of support (nodes) over time

The patient's illness initially defines why these health-bound networks exist. The patient or ego is unique as she or he serves as a core, as the pretext for its very existence. Nodes migrate in and out of closeness and nodes' roles with the patient around with the social support evolve. This form of complexity is a dynamic and continuous reorganization of the actors (Halley and Winkler 2008). People come. People go. Ideally, a person that comes, stays, helps, and influences positively is a node worth fussing over to stay invested.

Understanding the migration patterns will require policymakers to have a specific study period in mind (time x to time y). Direct migration means that a person moves from one sphere (and its network) to another with no intermediate steps to reach the final destination. A change of role from intimate engagement to no contact at all could be illustrated by a change in medical provider. There is no other support role to be served other than the clinical role. Is that clinician replaced in the network with one that will engage or engage more than the lost node? The nephrologist will not be taking the patient to the store. So the clinician skips to the polis (e.g., the patient is no longer a patient in the nephrology practice). This is a situation where replacement of nodes becomes important to support with policy. Social networks are organic in the sense that each thrives and is constrained by the social environment. Policy is tied beyond a nudge to the public. Networks affect choices and toggle with ethics. Members of a network self-organize (Halley and Winkler 2008). Choices and their resultant (un)ethical actions bolster or torpedo adherence and benefits to the collective.

Two important properties of complex networks proposed by Newman (2003) are specifically applicable here. First network resilience is concerned with maintaining the health of the network, via exploring the effects of the random removal of nodes from a network (Newman 2003). Second, community structure will encompass a stratification of the nature of support. This assortivity, as explained by Newman (2003), may be shown through exploring what determinants (e.g., race, age, nativity, and politics) are driving the formation of the levels of engagement of the nodes in a network. It is not assumed that every tie in a network would be designated as weak. Any transformation of composition of a network can be viewed as a reorganization or emergence of a complex network. Feld (1981) proposed that ties can tend to formulate around a shared purpose. The more investment, it becomes harder to leave even when the supporter is not in a good place.

> There must be one person…there must be at least a single common center—in this case, a specific person with whom an (node) pair commonly associates.
> (Louch 2000)

Intermediary migration could be an important indicator of systemic changes under a policy. What happens when a person changes intimacy and roles in connection with a patient? In terms of how this discovery may affect chronically ill patients, the reasons that caregivers may look forward in moving away from a supporting role could be a suggested policy issue. Does the policy account for reasons around caregiver network formation and fatigue in a way that uncovers the effects of homophily? Structural and emotional "closeness" brings to bear the need for understanding more fully the nature of direct and intermediary migration proposed by CMHN©.

A Kossinets and Watts (2009) model resulted in demonstrating that new network ties occur within people who already have friends in common which may perhaps to structural folds across these concentric networks. Outside of the digital world, socialization and connections happen where we live (see Battle-Fisher 2013; Allen 1977). The nodes must acknowledge that a link or tie exists. There must be an accounting on a political level that the emergence of new nodes is not created equal. Why does this inequality even matter? On a macrolevel, the microstructure becomes a motif for the divided realities of crossing insulated paths with people like ourselves and with a stake in it all.

Vedres and Stark (2010)'s idea of a structural fold is more recent and viewed in opposition to more established Burt's (1992) depiction of "structural holes" that was mentioned in the homeless teen analysis. In contrast to a weak tie, it is defined by its absence of intimacy and the infrequency of connecting (Hansen 1999). Vedres and Stark (2010) contended that in fact bridges supported by weakly invested ties are the least productive for cohesion within a network. Vedres and Stark (2010) explain passage of ideas within business organizations as dependent on intercohesion. With intercohesion, two or more interest groups could be influential in information flow. A new tie, as a structural fold, would replace the time and influence that a lost tie or actor may have supplied and bring in the strength of already being deeply tied across spheres.

It is hypothesized by this model that if members of the kidney network were not embedded into the other spheres, information about the extensive knowledge of the ESRD patient would not be diffused to people who were less involved and could help later. The structural folds are power brokers to combat "out of sight, out of mind." Now whether that information is truthful or even helpful is a question for another day. Whether that power is used for good or personal gain is another issue. If this becomes the case in this model, a strong tie may only remain effort intensive when called upon and the intercohesion may have a better chance of successfully supporting a network. The prior existence of a more intimate tie may, under the context of Jack's (2005) work, reactivate and become an active tie again. The point would be to work on maintaining ties with support persons in order for the tie to be reactivated in the future if the node becomes distal. Maintaining support networks does not just happen spontaneously and its inertia may work for or against health outcomes.

> There is a fundamental misconception to mistake for a "problem" what actually is only a mathematical "exercise." One would do well to remember the old Kantian maxim that experience without theory is blind, but theory without experience is a mere intellectual play. (von Bertalanffy 1962)

The Social Elastic Tie

Allegorically, for a moment, imagine a tie as an elastic band. Elastic bands bounce back into the shape even after it is stretched. A physical band possesses resilient properties, just like a network tie. When the two ends of the band are held taut, the force exerted between the two ends is tension. The band without an outside source pulling it will not exhibit tension. As such, in order for a node to be meaningful in most cases, there is a tie to another node. The exception to this condition is the isolate, one that is within the network boundary but unconnected to another node. Without delving into molecular, heat properties which play a huge role in the elastic's integrity, the elastic band that is overstretched always breaks off in the middle. The middle between the two nodes (rubber tie) is actually the cross section with the highest amount of tension. The length of the *social elastic tie* can be short or elongated. Recall that there are three characteristics to a tie: tie strength, intensity, and time duration (Borgatti and Halgin 2011). When the "elastic" tie begins to wear due in part to the characteristics just mentioned or also exogenous factors outside of a system, the resources or the will to stay connected may weaken the middle cross section of the tie between the two nodes. It is not the tension on the band pulling the ends apart that is the culprit toward physical breakage. The band breaks due to the amount of stress at its thinnest part. What is the point where fractures in the tie in social support can be avoided? How can policy help to circumvent social elastic tie failure, which breaks and then no longer exist?

As mentioned previously, the robustness of real networks is more real due to the exhibition of strong social organizational properties (Gao et al. 2012). If the network is larger, then that threshold may be more difficult to reach. It may be unrealistic to expect policy alone to increase substantially support networks large enough to tolerate such losses. However, if policy can serve to lessen the chance of losing people in the first place, that may be more plausible for policymakers to tackle. For instance, does a community have policies in place to increase support for caregivers to lessen the probability of burnout and fatigue? Is the policy leveraging what actually fights the burnout? In turn, adding the complexity of interdependent networks, loss of actors in any network can lead to a cascade of failures that reverberates across all networks (Gao et al. 2012). However, actor and tie failures are the main culprits to its overall stability. In other words, the realities of caregiving and living with the stress of ESRD may lead to losses of support ties and movements of support across spheres to a lesser intensive role.

There are exogenous parameters in a dynamic system insidiously at work, harkening Meadows' (1999, 2008) point of the often phantom nature of systemic feedbacks. I propose that social factors and stressors that affect social ties to supply valuable information for policymakers. While all of the following could possibly be framed as feedbacks, I ask that we take a moment to pay specific attention to the minutiae of a tie. What societal forces can create brittle ties that are so desperately needed for the sake of the patient? Mische (2011) wrote that to assure the relational durability within a network, the ties are those comprised of "shared history and values." Some time ago, the work of Rapoport (1953) said that we may only add to our friendships as we have access to. But often the biases become inexplicably tied to our life chances, those latent tentacles that push and pull on us often without us knowing. It does really become about who we know and who is giving up the goods, both as a structural and political phenomenon. From ordinary language to mathematical modeling, this is an interesting question for policy to acknowledge what hazards befalling chronically ill patients and their networks might be willing to shoulder.

References

Albert R, Jeong H, Barabasi A-L (2000) Attack and error tolerance of complex networks. Nature 406:378–382

Allen T (1977) Managing the flow of technology. MIT Press, Cambridge

Aristotle (1991) On rhetoric. A theory of civic discourse. Translated by Kennedy G. Oxford University Press, New York

Barnes JA (1974) Social networks. An Addison-Wesley module in anthropology. Addison-Wesley, Reading, pp 1–29

Battle-Fisher M (2013) Urban greenspce and collective health ownership. Mindful Nature 6(3):33–35. http://www.humansandnature.org/urban-greenspace-and-collective-health-ownership-article-159.php?issue=21. Accessed 14 April 2014

Borgatti S, Halgin D (2011) On network theory. Organ Sci 22(5):1168–1181

Burt R (1992) Structural holes: the social structure of competition. Harvard University Press, Boston

Burt R (2004) Structural holes and good ideas. Am J Sociol 110:349–399

Carpenter D, Esterling K, Lazer D (2003) The strength of strong ties—a model of contact making in policy networks with evidence from U.S. health politics. Ration Soc 15(4):411–440

Cooper A-J (1988) A voice from the south. Oxford University Press, New York

Cornwell B, Schumm L, Laumann E, Graber J (2009) Social networks in the NSHAP Study: rationale, measurement and preliminary findings. J Gerontol B Psychol Sci Soc 64B(Suppl 1): i47–55

Cornwell E, Waite L (2009). Social disconnectedness, perceived isolation and health among older adults. J Health Soc Behav 50(1):31–48

Department of Health and Human Services, Administration on Aging (n.d.) Projected future growth of the older population, population 85 and over by sex: 1900–2050. http://www.aoa.gov/Aging_Statistics/future_growth/future_growth.aspx. Accessed 12 June 2014

Everett M, Borgatti S (1998) Analyzing clique overlap. Connections 21(1):49–61

Feld S (1981) The focused organization of social ties. Am J Sociol 86(5):105–1035

Feld S, Carter W (1998) Foci of activities as changing contexts for friendship. In: Adams R, Allan G (eds) Placing friendship in context. Cambridge University Press, Cambridge

Gao J, Buldyrev S, Stanley H, Havlin S (2012) Networks formed from interdependent networks. Nat Phys. doi:10.1038/NPHYS2180

Granovetter M (1973) The strength of weak ties. Am J Psychol 78(9):1360–1380

Gregson J, Foerster S, Orr R, Jones L, Benedict J, Clark B, … Zotz K (2001) System, environmental, and policy changes: using the social-ecological model as a framework for evaluating nutrition education and social marketing programs with low-income audiences. J Nutr Educ 33(Suppl 1):4–15. doi:10.1016/S1499-4046(06)60065-1

Halley J, Winkler D (2008) Classification of emergence and its relationship to self-organization. Complexity 13:10–15

Hansen M (1999) The search-transfer problem: the role of weak ties in sharing knowledge across organizational subunits. Admin Sci Q 36:82–111

House JS, Landis KR, Umberson D (1988) Social relationships and health. Science 241:540–545

Jack S (2005) The role, use and activation of strong and weak network ties: a qualitative study. J Manag Stud 42(6):1233–1256

Kossinets G, Watts D (2009) Origins of homophily in an evolving social network. Am J Sociol 115(2):405–450

Long J, Cunningham F, Braithwaite J (2013) Bridges, brokers and boundary spanners in collaborative networks: a systematic review. BMC Health Serv Res 13:158

Louch H (2000) Personal network integration: transitivity and homophily in strong-tie relations. Soc Netw 22:45–64

Lynn J (2014) Strategies to ease the burden of family caregivers. JAMA 311(10):1021–1022

Meadows D (1999) Leverage points: places to intervene in a system. The Sustainability Institute, Hartland

Meadows D (2008) Thinking in systems: a primer. Chelsea Green, River Junction

Mische A (2011) Relational sociology, culture and agency. In: Scott J, Carrington P (eds) The SAGE handbook of social network analysis. Sage, London

Newman N (2003) The structure and function of complex networks. SIAM Review 45:167–256.

Palla G, Derenyi I, Frakas I, Vicsek T (2005) Uncovering the overlapping community structure of complex networks in nature and society. Nature. doi:10.1038/nature03607

Rapoport A (1953) A marginalized network highlights both the mathematical finiteness as well as socio-political constriction on life chances. Bull Math Biophys 15:523–533

Rosenthal M (2014) Caregiver-centered care. JAMA 311(10):1015–1016

Sterman J (2002) All models are wrong: reflections on becoming a systems scientist. Syst Dyn Rev 18:501–531

Tamura MK, Tan J, O'Hare A (2012) Optimizing renal replacement therapy in older adults—a framework for making individualized decisions. Kidney Int 82(2):261–269

United States Renal Data System (2013) USRDS 2013 annual data report: atlas of chronic kidney disease and end-stage renal disease in the United States. National Institutes of Health, National Institute of Diabetes and Digestive and Kidney Diseases, Bethesda

Vedres B, Stark D (2010) Structural folds: generative disruption in overlapping groups. Am J Sociol 1115(4):1150–1190

von Bertalanffy L (1962) General system theory—a critical review. Gen Syst 7:1–20

von Bertalanffy L (1968) General system theory—foundations, development, applications. Braziller, New York

White D, Houseman M (2002) Navigability of strong ties: small worlds, tie strength and network topology. Complexity 8(1):72–81

Index

Printed in the United States
By Bookmasters